Early Astronomy

SPRINGER
STUDY
EDITION

Springer
New York
Berlin
Heidelberg
Barcelona
Budapest
Hong Kong
London
Milan
Paris
Santa Clara
Singapore
Tokyo

Hugh Thurston

Early Astronomy

With 139 Illustrations

Springer

Hugh Thurston
University of British Columbia
Department of Mathematics
121—1984 Mathematics Road
Vancouver, BC
Canada V6T 1Y4

All photographs first appeared in *The Griffith Observer* (used by permission)

Cover illustrations: Tycho's great sextant (Figure 10.6) and the
Systema Tychonicum (from *The Scientific Image*)

Library of Congress Cataloging-in-Publication Data
Thurston, Hugh, 1922–
  Early astronomy/Hugh Thurston,—1st ed.
    p. cm.
  Includes bibliographical references and index.
  ISBN 0-387-94107-X.—ISBN 3-540-94107-X
  1. Astronomy. Ancient. I. Title.
QB16.T48 1993
520′.93—dc20   93-5142

Printed on acid-free paper.

First softcover printing, 1996.

Production coordinated by Brian Howe and managed by Bill Imbornoni; manufacturing
supervised by Genieve Shaw.
Typeset by Best-set Typesetter Ltd., Hong Kong.
Printed and bound by Edwards Brothers, Inc., Ann Arbor, MI.
Printed in the United States of America.

9 8 7 6 5 4 3 2 1

ISBN 0-387-94107-X Springer-Verlag New York Berlin Heidelberg (hard)
ISBN 3-540-94107-X Springer-Verlag Berlin Heidelberg New York (hard)
ISBN 0-387-94822-8 Springer-Verlag New York Berlin Heidelberg SPIN 10541595 (soft)

Contents

Contents

Introduction

People must have watched the skies from time immemorial. Human beings have always shown intellectual curiosity in abundance, and before the invention of modern distractions people had more time—and more mental energy—to devote to stargazing than we have. Megaliths, Chinese oracle bones, Babylonian clay tablets, and Mayan glyphs all yield evidence of early peoples' interest in the skies.

To understand early astronomy we need to be familiar with various phenomena that could—and still can—be seen in the sky. For instance, it seems that some early people were interested in the points on the horizon where the moon rises or sets and marked the directions of these points with megaliths. These directions go through a complicated cycle—much more complicated than the cycle of the phases of the moon from new to full and back to new, and more complicated than the cycle of the rising and setting directions of the sun. Other peoples were interested in the irregular motions of the planets and in the way in which the times of rising of the various stars varied through the year, so we need to know about these phenomena, i.e., about retrogression and about heliacal rising, to use the technical terms. The book opens with an explanation of these matters.

Early astronomers did more than just gaze in awe at the heavenly bodies; they tried to understand the complex details of their movements. By 300 B.C. the Babylonians had devised an intricate arithmetical system for this purpose, and by A.D. 150 the Greeks had developed a powerful geometrical theory that was capable of representing the motions of the sun and planets remarkably accurately.

Another great early civilization, Egypt, had disappointingly little astronomy, but China's contribution was substantial and interestingly different from that in the west. It reached its peak in the late thirteenth century, to be superseded soon after by western astronomy introduced by Jesuit missionaries. China's astronomy had a nudge from Babylon in early historical times, and classical Indian astronomy almost certainly developed from early Greek geometrical theories, but the Mayas, far away in central America, developed an astronomy entirely independent of the astronomy of the Old World.

These astronomies all had the sun moving round the earth. They were superseded eventually by Copernican astronomy, in which the earth moves round the sun.

Copernicus led to Kepler, the first person to find the true shape of the planets' orbits. And then—some forty centuries after the megaliths— came a natural watershed. Up to this date, astronomy had been a matter of patient observation with simple instruments, little more than jointed rods and pivoted rings, followed by pencil-and-paper calculations. But Galileo, a contemporary of Kepler, altered all this. First, he used the newly-invented telescope to look at the moon and the planets. This revolutionized astronomical observation, leading eventually to today's sophisticated observatories with their computer-controlled telescopes. Second, he began to develop a theory of motion. In the hands of Isaac Newton such a theory would eventually explain the celestial motions in terms of well-founded physical principles. Astronomy became part of physics.

This book covers astronomy from the beginning up to the time of Kepler. Most of this was developed at a time when other sciences, notably physics and chemistry, had scarcely started. The scientific edifice built up over this period is one of the triumphs of human intellect.

I would like to acknowledge the willing and friendly help of Mr. S.Y. Tse, of the Asian Centre Library, University of British Columbia, of Professor Ashok Aklujkar, of the Department of Asian Studies, University of British Columbia, and of Dr. Edwin C. Krupp, of the Griffith Observatory, Los Angeles.

CHAPTER 1

Early Stargazers

The Celestial Bowl

If we look at the sky on a fine night—preferably from a ship at sea or a place deep in the countryside, not from among the bright lights of a town—we seem to see the inside of a dark bowl, studded with points of light. You and I, living in the twentieth century A.D., know that this is an illusion. There is no bowl, and the stars are at different distances from us: some a few light-years away, others many hundreds. But the illusion is strong: Chinese, Greek, Arab, and medieval European astronomers all treated the stars as lying on a sphere around the earth.

How big is this celestial bowl? Certainly out of reach. Not just hundreds or thousands of yards away, but many miles. If we make long journeys, the pattern we see in the sky does not change in the slightest. If we climb the highest mountains, the stars seem no nearer. Ptolemy (second century A.D.), in his astronomical compendium, the *Almagest*, said:

the earth is essentially a point compared with its distance from the sphere of the so-called fixed stars,

which amounts to saying that the radius of the celestial bowl is essentially infinite. However, in a later but not-so-well-known work (*Hypotheseis ton Planomenon*) he gave the radius of the celestial bowl as 19,865 times the radius of the earth. We will see on page 172 how he reached this extraordinary figure.

Copernicus (fifteenth century A.D.) agreed with Ptolemy's first thoughts. He wrote:

the heavens are infinitely large compared with the earth, and it is by no means clear how far their immensity reaches.

The modern point of view is as follows. For certain purposes we are interested only in the directions, not the distances, of the sun, moon, stars, and planets. Observations and calculations by navigators are concerned only with directions, and the same applies to calculations used in astrology. For such calculations, we assume that the celestial bodies lie on an imaginary sphere of arbitrary radius, called the celestial sphere, and we apply the techniques of spherical trigonometry to this sphere. So although we know that the sun is nearer than, say, Sirius, we allow ourselves to ignore this fact in contexts where it does not matter. The modern "celestial sphere" is a mathematical abstraction, but agrees well with the naive view that the heavens look like a sphere.

One point about spheres needs to be explained. If we slice a well-shaped orange with a knife, the cut is circular; in other words, a section of a sphere by a plane is a circle. The nearer the plane is to the center of the sphere, the bigger the circle. The biggest possible circle is obtained when the plane passes through the center of the sphere and it is cut exactly in half. Such a circle is called a *great circle* and its radius is the same as the radius of the sphere. Great circles are important in astronomy because if a heavenly body is moving in a plane through the earth, its apparent path on the celestial sphere, as seen from the earth, is a great circle.

The Constellations

The stars are scattered over the sky pretty well at random, but to an imaginative stargazer they form pictures [1]. The group of bright stars shown in Figure 1.1(a) seems to some people to form a ladle. This constellation is, in fact, the "big dipper." Other people see this group as a wagon and call it "Charles's wain." Yet others see it as the "great bear."

All the world's peoples seem to have formed pictures in the sky. To the Chukchi (a Siberian tribe) our Orion is an archer and our Pleiades are his target. Our Leo is his wife. Six stars of our Great Bear are slingers; the seventh is a fox. Gemini are two elks; Cassiopeia is five reindeer. To the Eskimo, our Great Bear is a reindeer and Cygnus is three kayaks. To the Tlingit (on the Pacific coast of Canada) the Pleiades are a sculpin (a common fish). To the Baikari (in Brazil), our Orion is a manioc-drying frame; Pisces and Argo together form a heron. To the Hottentots, Orion's belt and sword are six zebras. The Maoris saw a legendary canoe where we see Orion's belt, the Pleiades, and the Southern Cross. To the Bushmen, our Gemini are two antelopes.

Our constellations were mostly taken over from the classical Greeks, though we use the later Latin names for them. The southernmost regions of the sky are not visible from Greece or the parts of the world to which the Greeks traveled, so later astronomers devised constellations to fill the gap. That is why some southern constellations have unexpectedly modern names, like "telescope."

The Greeks did not start entirely from scratch, but took over some of the constellations from the Babylonians. For example, they took over the

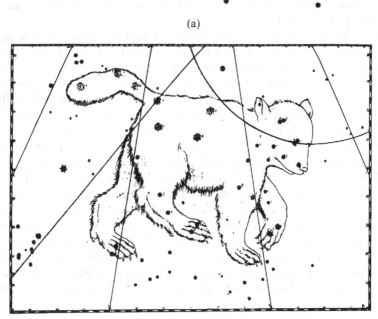

(a)

The Great Bear. From Bayer's *Uranometria* (1603).

(b)

FIGURE 1.1.

constellation *gir* (which means scorpion) under the name *skorpios*. They took over, too, the scales and the lion. But other Greek constellations were quite different from those of the Babylonians. The Babylonian *iku* [field], for instance, consists of four stars; three are part of the Greek constellation *hippos* [horse—modern name, Pegasus] and one is part of Andromeda [2].

One use of constellations is for describing positions in the night sky, by saying, for instance, "the moon is midway between the two leftmost stars of Pisces." Were the constellations devised for this purpose? E.W. Maunder, when he investigated astronomy in biblical times, found evidence that the Greek constellations were no mere pictures: Hydra, for example, is a long thin constellation made up of rather dull stars which no one, idly imagining pictures in the sky, would be tempted to visualize as anything in particular; but it lies in just the right position to play a useful role in a system of coordinates [3]. His suggestions were pushed further by Michael Ovenden (see page 135 below—we need to study how the positions of stars changed over the ages before we can go into details of Maunder's theories).

The Chinese had a list of twenty-eight constellations, known as *xiu*, that between them completely encircle the heavens and were used to describe positions in the sky.

There are two mysteries about the constellation whose principal stars are shown in Figure 1.1(a). It is no surprise that anyone should think it looks like a pan, and indeed a common name for it is "the big dipper," while the Chinese call it *Bei tou* [Northern ladle]. Nor are we surprised to find it called "the plough" or "Charles's wain." It is more surprising that the Greeks called it a bear, but the Greeks did not set much store by realism, and peopled the sky with hunting dogs and a horse as well as mythological characters. The surprise comes when we find that some North American Indian tribes, although entirely isolated from the classical world, also called it a bear. In particular, the Micmacs called the Great Bear *Mouhinne* [bear] and the Little Bear *Mouhinchiche* (diminutive of *Mouhinne*). A second surprise comes when we notice that, even though the Greeks were familiar with bears, the Greek bear has a long tail: Figure 1.1(b) shows a typical picture of the constellation based on the Greek description. American aborigines were more sensible. The Algonkins and Iroquois, for example, called the constellation "The bear and the hunters," the hunters being the three stars that the Greeks called the tail [4].

The Rotation of the Heavens

The shapes of the constellations are fixed. They do not change like the pattern made by a flock of starlings against the sky. But their positions are not fixed: *the complete bowl rotates about a fixed point once a day.* This point is called the celestial pole.

Today we are used to thinking of the earth as spinning and the sky as still, but early astronomers, whether Babylonians, Egyptians, or Chinese, never even considered that it might be the earth that was moving. The Greeks did consider the possibility but rejected it (see page 121). The modern logically strict point of view is that all motion is relative: to say that the sky is turning clockwise relative to the earth is the same as saying that the earth is turning anticlockwise relative to the sky. To understand early astronomy, it is best to take the same point of view as the early astronomers. After all, this is the natural point of view—even astronomers say "I saw the sun set," not "I saw the horizon rise and cover the sun."

The celestial pole is not directly overhead. If it were, the stars would move horizontally. As it is, they rise and set, or at any rate some do—some are close enough to the celestial pole for the circles on which they move to be entirely above the horizon. Consequently, as the heavens rotate we see altogether more than a hemisphere, and from quite early times nearly all astronomers regarded the heavens as a complete sphere, though a Chinese theory, the *Gaitian* [Celestial lid] theory, in which the heavens are less than a sphere, lingered on until A.D. 200 (see page 90).

If we are in the northern hemisphere, the celestial pole in the part of the celestial sphere visible to us is the *north* pole. I will always describe the heavens from the point of view of someone in the northern hemisphere—most, if not all, of the important early astronomers were Northerners.

Anyone who would like something more concrete than a description in words of the celestial motions, but who has no access to a planetarium, can make a miniplanetarium out of a bubble-umbrella. Most bubble-umbrellas are roughly hemispherical; let us suppose that ours is a perfect hemisphere. It represents half of the celestial sphere—the northern half. The southern half we have to imagine. The center of the complete sphere is a point on a shaft of the umbrella; let us call this point O. Set the open umbrella on a stand with the shaft pointing toward the celestial pole. Finally, glue tinsel stars to the canopy of the umbrella to form a map of the constellations of the northern hemisphere.

Now imagine a small sphere, perhaps a millimeter across, at the point O. This is the earth. An astronomer—quite microscopic on this scale—standing on it can see only half the sky at any one time; the earth blocks his or her view of the other half. To get this effect, make a circular table a trifle smaller than the umbrella in diameter, cut a small hole in the middle, and put the shaft of the umbrella through the hole, with the point O on a level with the table. The set-up is like a garden-table-with-sunshade, except that the shaft of the umbrella is not vertical but is pointing to the celestial pole. The rim of the table represents the horizon. If we rotate the umbrella, the tinsel stars will reproduce the motion of the stars in the sky.

A star close to the pole, such as the Pole Star, which is only 1° away, moves in a small circle around the pole; a star further away moves in a larger circle; and a star far enough from the pole moves in a circle that dips below the horizon, and will rise and set. If we are in latitude 50° the celestial pole is 50° above our horizon, and a star 50° from the pole just touches our horizon at its northernmost point as the celestial sphere turns. Stars nearer the pole will not set. A star a little more than 50° from the pole sets briefly a little to the west of north and rises the same distance east of north. A star further from the pole dips further below the horizon, setting further west and rising further east. A star on the celestial equator sets due west and rises due east. Stars south of the celestial equator rise south of east and set south of west provided that their angular distances from the south pole are not less than 50°; a star nearer the south pole than 50° will never rise but will remain forever invisible to an observer at latitude 50° north.

Here, for latitude 50°, are rising-azimuths corresponding to various distances from the pole. (An azimuth is a horizontal direction measured clockwise from north. Thus azimuth 45° is north-east.)

Angle from pole	90°	85°	80°	75°	70°	65°	60°	55°	51°	50°
Azimuth	90°	82°	74°	66°	58°	49°	39°	27°	12°	0°

At a different latitude the correspondence will be different:

Latitude 55°

Angle from pole	90°	85°	80°	75°	70°	65°	60°	55°
Azimuth	90°	81°	72°	63°	53°	42°	29°	0°

Latitude 35°

Angle from pole	90°	80°	70°	60°	50°	40°	35°
Azimuth	90°	78°	65°	52°	38°	21°	0°

These figures are for a flat horizon and an observer at ground level. If the height of the horizon is known it is easy to correct for it, and it is equally easy to correct for the height of the observer. However, there is one thing that is not easy to correct for, namely *refraction*, the bending of light-rays as they pass through the earth's atmosphere. Its effect is to make stars near the horizon seem higher than they actually are. At the moment when you see the last gleam of the sun at sunset, the sun is actually below the horizon—a geometrically straight line from your eye to the sun would not clear the horizon, and the light from the sun is reaching your eye via a curved path. Tables of refraction have been drawn up, but no one is sure how accurate they are, and in any case the effect of refraction varies with temperature and humidity. Refraction can easily make a difference of half a degree.

There is one other correction that should be mentioned here: the correction due to "parallax." To say that the moon goes round the earth means, precisely speaking, that the moon goes round the center of the earth. But we are not at the center of the earth, we are somewhere on the earth's surface. Therefore, the direction from us to the moon is not quite the same as the direction from the center of the celestial sphere to the moon. The difference is called the moon's *parallax*. In Figure 1.2 the circle represents the earth with center C. M is the (center of the) moon. The diagram is not to scale: the distance from the earth to the moon is

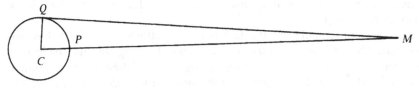

FIGURE 1.2.

actually about sixty times the radius of the earth. If we happen to be at P, so that the moon is directly overhead, the parallax is zero, because PM and CM are in the same direction. If we are at Q, where MQ just touches the earth, so that the moon is on our horizon, as it would be at rising or setting, the parallax is at a maximum: it is, of course, the angle CMQ. There is a slight complication because the distance of the moon from the earth varies slightly, and so the parallax of the rising or setting moon is not always the same. In fact, it varies from 0.898° to 1.025°. Fortunately, the sun, the planets, and the stars are at such vast distances from the earth that their parallaxes are quite negligible.

Instead of the angle from the pole we could use the angle from the equator, the technical term for which is *declination*. Astronomers have to get used to the fact that whereas on earth the angle from the equator is latitude, in the sky the angle from the celestial equator is not celestial latitude. There is such a thing as celestial latitude, but it is something different, as we will see on page 32. Declinations north of the equator are positive; south, negative. The relation between the declination and the angle from the pole is simple: a declination of x degrees corresponds to an angle of $90 - x$ degrees from the north pole.

Small Irregularities

I said earlier (page 4) that the fixed pattern of the constellations rotated once a day about a fixed point. There are three things wrong with this statement:

(1) the center of rotation is not fixed;
(2) the pattern is not fixed; and
(3) one rotation does not take exactly one day.

However, the statement is very nearly correct: the center of rotation (the celestial pole) moves very slowly, the pattern is mostly fixed, and the period of rotation is very nearly one day. This is a striking example of something that we will meet time and time again: we continually find general regularities tempered by small irregularities. Investigation of the small irregularities has been vital to the progress of astronomy.

The first irregularity, the movement of the pole, called *precession*, is so slow that it is hard to detect; neither the Egyptians nor the Babylonians knew about it. It was first discovered about 150 B.C. by the Greek astronomer Hipparchus.

The second irregularity, the fact that the pattern is not fixed, has two aspects. One is that the stars are not gems studding the underside of a heavenly bowl, but bodies floating through space. They do move relative to each other and the shapes of the constellations do change, but the changes in position as seen from the earth are tiny even over a time-span of thousands of years. This motion is known to modern astronomers as the *proper* motion of the stars, and was not discovered until A.D. 1718. It therefore does not concern us here. But the other aspect of this irregularity does. There are a number of apparently starlike bodies that move very obviously against the fixed background. The Greeks called them *planetes* (wanderers) and we call them planets.

The third irregularity is a vital one: the period of rotation of the constellations around the earth is not exactly one day. What do we mean by one day? It is the time taken for the sun to revolve once around the earth. Therefore the constellations do not revolve around the earth at quite the same speed as the sun.

The Sun

Perhaps in very early times the study of the sun was separate from the study of the stars: stargazing at night, sun-worshipping by day. But quite soon people must have realized that the stars are there all the time; they are not created at dusk and destroyed at dawn, but are merely swamped by the light of the sun. The sun, too, is not a new fireball each morning, but the same one revolving round and round the earth. People just beginning to grasp this idea had some fascinating myths "explaining" how the sun was transported underground from the place where it set one evening to the place where it rose the next morning.

The illusion of the hemispherical bowl is not as striking for the empty blue sky by day as for the star-studded sky at night. Nevertheless, most people realized that the sun revolved with the heavens round the earth. But not quite. Not only is the period of rotation not quite the same, but for people outside the tropics the sun is higher in the sky in summer. No one can miss the difference between a typical winter day and a typical summer day in Greece, Babylonia, or China. And although the Egyptians, whose climate is more tropical, did not experience the striking alternation of cold winter with warm summer, they did have one vitally important annual event, namely, the flooding of the Nile caused by the monsoon rains in the mountains where the Blue Nile rises.

The time when the sun is highest is called midsummer. We will see (pages 10, 57, and 97) how it can be found. If we count the days between one midsummer and the next, we will find that there are 365 as a rule, though occasionally we might make the answer 366. If we count over a long period, we will find an average pretty close to $365\frac{1}{4}$. The interval from one midsummer to the next (or, for that matter, one midwinter to the next) is called a *year*.

Now let us think a little more about the apparent rotations of the sun and of the stars around the earth. We have seen that their periods are not quite the same. How different are they? And how was the difference discovered? Even before clocks were invented, the stargazer could judge reasonably accurately the time halfway between sunset and sunrise—i.e., midnight—and tell that from one midnight to the next the stars made a little more than one revolution, because after a couple of weeks the patterns of the constellations at midnight had advanced noticeably. The pattern takes a year to advance through a whole revolution.

There are therefore two periods that we could call a "day." One is the time in which the sun makes one revolution round the earth. We call this the solar day. The other is the time in which the celestial sphere makes one revolution. We call this the sidereal day. In everyday life, daylight is important; therefore the legal, calendrical, day is the solar day.

One revolution per year is not quite 1° per day. The heavens rotate through 360° per day, so 1° takes four minutes. Therefore a sidereal day is not quite four minutes shorter than a solar day.

The fact that the sun is revolving round the earth more slowly than the celestial sphere means that it is not fixed on the sphere: it moves relative to the "fixed" stars. The celestial sphere rotates westward; therefore the motion of the sun relative to the sphere is eastward. It takes a year to travel completely round the celestial sphere.

The fact that the sun is higher in the sky in summer than in winter means that its movement round the celestial sphere is oblique. By noting the constellations that appear on the horizon where the sun disappears at sunset, and the constellations that disappear where the sun appears at sunrise, we can get some idea of the path of the sun on the celestial sphere. It is a great circle, tilted at an angle to the equator. We will see on page 44 how this angle can be measured; it is now about 23.45°. It has been changing slowly over the centuries. In prehistoric times (3000 B.C.) it was about 24.04°. In classical times (300 B.C.) it was about 23.74° [5].

The sun takes a year to go round its orbit. It is at the northernmost point in the (northern hemisphere's) summer, crosses the equator in the autumn, reaches the southernmost point in the winter, and crosses back to the northern half of the celestial sphere in the spring. The ancient Chinese likened the motion of the sun to that of an ant crawling on a millstone [6].

The Directions of Sunrise and Sunset

As the sun moves round its orbit its angle from the pole varies. Therefore the point on the horizon where it rises will change—it is only roughly true that the sun rises in the east and sets in the west. At latitude 50° north we would see it rise at azimuth 50° (roughly north-east) at midsummer, and at azimuth 130° at midwinter: a swing of 80°. In higher latitudes there will be a wider swing.

It is fascinating to follow the rising- and setting-directions of the sun. I once lived for some months in a house whose verandah faced west over a flat plain. Ridges, notches, and hillocks enabled me to pick out the points on the horizon where the sun set. I could not help noticing that as time passed from August to December, the sun set a little further to the left each day. Had I lived there longer I could have followed the cycle of sunset from furthest right at midsummer to furthest left at midwinter and back, and noticed that the cycle repeated itself exactly from year to year. I could have found the furthest left and furthest right directions of sunset and made a note of the appropriate marks on the horizon if there happened to be any conveniently placed. Or I could have set up posts as markers, a "backsight" on the verandah and two "foresights" a convenient distance away.

In spring and autumn the setting-direction changes fairly rapidly, about a degree per day. (This may not seem much until we realize that it is twice the sun's apparent diameter.) It slows up considerably as it reaches the midsummer and midwinter directions. This is rather like the pendulum of a grandfather clock, which moves fastest in the middle of its swing and seems to hover motionless for an instant at the end of each swing while changing direction. (Anyone who has seen a Foucault's pendulum swinging will have found the slowing-down at the ends of the swing very obvious.) Because the setting-direction of the sun seems to remain momentarily unchanging at these times, the instants when the sun is at its greatest northerly and southerly declinations are called *solstices*, from the Latin *solstitium*, literally a "stand-still of the sun." The points where the sun is at these times, i.e., the most northerly and southerly points of its orbit, are also called solstices, so that the same word "solstice" can mean either a point in space or a point in time. The same applies to the word "equinox": an equinox is either a point where the sun's orbit crosses the equator or an instant when the sun crosses the equator. When the sun is at an equinox it spends half its time below the horizon, and so day and night are then equally long.

The setting-directions and rising-directions of the sun could serve as a calendar, and people in various parts of the world have used them for this purpose. The Zuñi, in North America, celebrated midwinter when the rays of the rising sun struck a certain point on a certain mountain: they described the solstice as the time when the sun set for four successive days at the same point. The Hopi dated not only celebrations but also seasonal activities such as sowing and reaping by the position of the sun. This kind of sun watching was also practiced in Melanesia, in Africa— where the Zulu said that when the sun moved from its winter position it was "going to fetch the summer"—and even in Norway, where the peasants celebrated St. Paul's day, Candlemas, etc., on the dates when the sun rose or set over certain peaks. In Sarawak, there are two oblong stones that point to a hill; it is time to sow when the sun sets behind the hill. At Batu Salu, a block of stone has a hollow worn in its top surface; priestesses used to sit in this hollow to watch the sun set behind a certain peak, in order to determine the time for sowing [7].

Besides evidence from anthropologists about existing peoples, there is evidence from archaeologists that early peoples marked the directions of sunrise. There are temples in Egypt, Guatemala, and Mexico marking these directions [8]. There is a barrow at Newgrange in Ireland (dated about 3000 B.C.) so constructed that at midwinter sunrise the rays of the sun shine through an opening over the doorway onto a plinth at the back of the interior chamber [9], see Figure 1.3. Above all, there is Stonehenge (which will be described on page 45 onward).

The direction of moonrise also changes, of course. Although no existing primitive tribes follow the directions of moonrise, there is strong evidence at Stonehenge and in western Scotland that early peoples did so. When we turn to the stars the archaeological evidence is weak. Existing primitive tribes have shown little interest in the directions in which stars rise and set, but we can quote two examples. The Ammasalik Eskimos made spring begin when the sun rose at the same point of the horizon as the star Altair. Polynesian and Micronesian navigators used stars to point their canoes in the right direction in their long voyages across the Pacific [10]. It is when a star is near the horizon that it is useful for marking a direction—a star directly overhead is useless—and on a long voyage a navigator would follow a succession of stars rising or setting at the same point of the horizon. There is little evidence that anyone followed the rising- or setting-directions of any of the planets, except that the Maya may have marked the southernmost rising of Venus (see page 23) [11].

The Irregular Sun

Let us suppose that we can find the time of the solstices correct to the nearest day. If we count the days, we find something unexpected: the time from midsummer to midwinter is not the same as from midwinter to midsummer. Why should this be? The change in the sun's declination is caused by the movement of the sun round its orbit. If the sun moved uniformly round the celestial sphere it would take exactly half a year to go from one solstice to the other, because they are diametrically opposite each other. Therefore the sun must move irregularly.

The interval from midsummer to midwinter is $184\frac{1}{2}$ days as against $180\frac{3}{4}$ from midwinter to midsummer, so the difference is not great. The two figures vary as the centuries pass: in 180 B.C. they were $180\frac{1}{2}$ and $184\frac{3}{4}$, respectively [12].

FIGURE 1.3. The opening over the door at Newgrange which allows sunlight to shine down the passage at midwinter sunrise. (Photograph Robin Rector Krupp, Griffith Observatory.)

We have here one more example of the theme that runs through astronomy—a general regularity tempered by small irregularities.

The Moon

The most obvious thing about the moon is that it changes shape: from nothing to a thin crescent to a gibbous shape to a circular disk, and back again. What we see is the part lit by the sun: the shapes are precisely the shapes we see if we shine a light on a ball and view it from various angles; and the lighted part of the moon is always facing the sun.

The time from one complete disappearance to the next is about $29\frac{1}{2}$ days. This period is the *month*. To distinguish it from other months (such as the 30-days-hath-September calendar month) we may call it the *synodic* month.

The moon revolves around the earth with the celestial sphere—but again, not quite. It slips eastward against the background pattern, that is, in the same direction as the sun, but over twelve times as fast, making a complete revolution in about 27 days. If we watch the moon for a long period against the background of the stars, we find that there is a great circle round the celestial sphere that the moon never strays far from. Moreover, eclipses of the moon occur only when the moon is exactly on this circle, which is therefore called the ecliptic. When the moon is eclipsed, its direction from the earth is opposite to that of the sun, and the fact that the moon is always full when it is eclipsed confirms this. The cause of an eclipse is now obvious: it is the shadow of the earth. And the ecliptic must be the path of the sun round the celestial sphere. If the moon also moved precisely along the ecliptic it would be eclipsed every month at full moon, but in fact it diverges up to 5° from the ecliptic.

There are two important technical terms used to describe the position of a body which, like the moon, moves close to the ecliptic. Let X be the point of the ecliptic nearest to the body. When X coincides with the sun, the body is said to be in *conjunction*. When X is at the opposite point of the ecliptic from the sun, the body is in *opposition*. Thus at full moon the moon is in opposition. If the moon happens to be right on the ecliptic when it is in opposition, it will be eclipsed then.

The Directions of Moonrise

The easiest way to follow the complicated changes in the directions of moonrise (and moonset) is to start from the motion of the sun on the celestial sphere. The moon's orbit does not quite lie in the plane of the ecliptic, but is at an angle of about 5° to it. In other words, the moon's orbit is at an angle of 5° to a plane which is itself at an angle of about $23\frac{1}{2}°$ to the plane of the equator. What, then, is the angle between the moon's orbit and the plane of the equator? Clearly, it cannot be more than $23\frac{1}{2}$ + 5°, nor less than $23\frac{1}{2}$ − 5°. In fact, it varies continuously between these two values. (More precise figures are 28.6° and 18.3°.) The reason is that although the angle of tilt stays the same, the direction of tilt changes.

The way in which the direction of tilt changes can be visualized as follows. Imagine that the positions of the sun and the moon, as seen from the earth, are projected onto the celestial sphere. The sun's orbit will be a great circle at an angle of $23\frac{1}{2}°$ to the equator, crossing it at the equinox points. The moon's orbit will be a great circle at an angle of 5° to the sun's orbit, and will cross it at two diametrically opposite points. These points are called *nodes*. As the direction of tilt varies the nodes move

Perspective   Side view

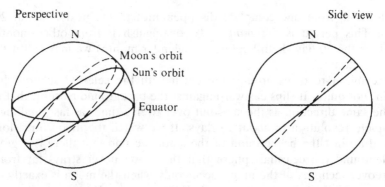

Moon's orbit at maximum angle to equator; nodes at equinox points.

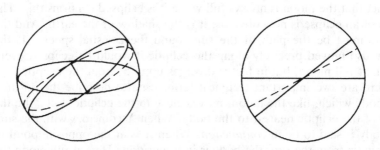

Moon's orbit at minimum angle to equator; nodes at equinox points.

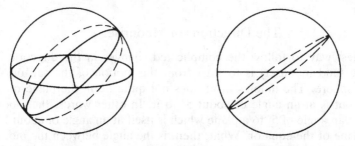

Halfway between the above situations; nodes at solstice points.

FIGURE 1.4.

round the sun's orbit. Figure 1.4 shows three positions of the nodes. In fact, the nodes move round the sun's orbit at a fairly constant rate, taking 18.6 years to go once round. This motion is called "the regression of the nodes."

The moon takes, on average, 27.32 days to go once round its orbit.

The effect of a 5.25° difference between the declinations of the sun and moon on the directions in which they rise, at latitude 50°.

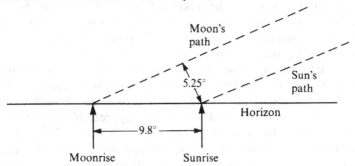

Sun's greatest declination, 23.35°; azimuth, 51.75°.
Moon's greatest declination, 28.60°; azimuth, 41.87°.

FIGURE 1.5.

This period is less than a synodic month, because a synodic month is the time the moon takes to go once round its orbit with respect to the sun. At full moon, for instance, the moon is opposite the sun. 27.32 days later it will be back in the same position, but it will then not be opposite the sun, because in those 27.32 days the sun has moved on a little. It takes the moon another couple of days to catch up with the sun; not until then will the moon be full again, and not until then will a month have passed since the last full moon.

In the 27.32-day period, the moon's declination varies from $28\frac{1}{2}°$ to $-28\frac{1}{2}°$ if the orbit is at $28\frac{1}{2}°$ to the equator; it varies from $18\frac{1}{2}°$ to $-18\frac{1}{2}°$ if the orbit is at $18\frac{1}{2}°$ to the equator, and so on. The difference in declination between the sun and the moon will make them rise at different points on the horizon. (See Figure 1.5)

The Cycles of the Moon

What does a moon-watcher actually see? Let us suppose that we start watching at the time of full moon. We will see the moon rise at sunset in a direction roughly opposite to that of the setting sun. The next day it will rise nearly an hour after sunset and will not be quite full. Each succeeding day the moon will rise later and wane more. Half a month after full moon is the time of new moon. A day or two before this the moon will have become too thin a crescent too near the sun to be seen.

Here is perhaps the place to explain that to an astronomer "new moon" is the instant when the moon passes the sun as they go round in their orbits, not the first instant after this when the moon becomes visible again, as in the popular superstition "turn your money over when you see the new moon."

The day after new moon the moon will rise and set a little later than the sun, and this day or the next will be visible just before it sets. (This is when the loose change gets turned over.) Moonset is nearly an hour later each day, and the moon continues to wax until it is full.

The azimuth of moonrise depends on the moon's declination. As explained above, the moon's greatest declination is about $28\frac{1}{2}°$, and occurs when the nodes are in the equinox positions (Figure 1.4, top). The positions of the nodes will not change appreciably in a month, so in a month its declination will change from $28\frac{1}{2}°$ to $-28\frac{1}{2}°$ and back. The declination of moonrise will not change quite so much, unless it happens that the moon is rising at the instant when it reaches its greatest declination, and again at the instant when it reaches its least declination. If, however, it reaches its greatest declination halfway between two moonrises, then at each of these moonrises the declination will be about 0.2° short of the maximum. For a latitude of 50° the azimuths corresponding to $\pm28\frac{1}{2}°$ are 138° and 42°. Thus the moonrise will vary between these azimuths (or a trifle less unless the extreme declinations both occur at moonrise)—a swing of 96°.

About 9 years later, the nodes will have gone halfway round the equator (Figure 1.4, middle) and the moon's declination will vary in one month only from $-18\frac{1}{2}°$ to $18\frac{1}{2}°$. The azimuth of moonrise will then vary only between 118° and 62° (or a trifle less)—a swing of only 56°.

We cannot actually watch the moon rise for a whole month because for half of the month the moon will rise in daylight. However, we can fill in the gap by watching the moon set. The moonrise azimuth on one day is opposite to the azimuth halfway between the moonset azimuths the day before and the day after.

Watching the direction of moonrise for a month is like watching the direction of sunrise for a year but with an important difference: the amount of swing is not always the same but varies gradually from its minimum to its maximum (at latitude 50°, from 56° to 96°) and back.

For anyone following the rising of the sun and the moon, there are six particularly interesting azimuths. Four of these are obvious: the azimuths of the northernmost and southernmost sunrise and moonrise. Two are not so obvious: the extreme azimuths of moonrise in those years when the swing is least. For latitude 50° and a flat horizon these azimuths are pictured in Figure 1.6, in which the moonrise azimuths are labeled M_1 to M_4. For a higher latitude the picture would be similar but the angles would be larger; for a lower latitude they would be smaller. For a hilly horizon the picture could, however, be appreciably distorted.

There is a small extra complication that should be mentioned: the angle between the moon's orbit and the sun's orbit is not quite fixed, but varies periodically between 5.0° and 5.3°, being at its greatest when the sun

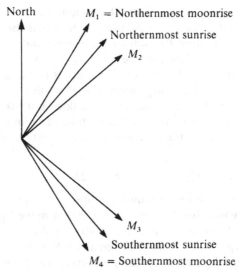

North M_1 = Northernmost moonrise

             Northernmost sunrise

                M_2

                  M_3

             Southernmost sunrise

           M_4 = Southernmost moonrise

M_2 and M_3 show the northernmost and southernmost moonrise when the
swing is least. The angles are roughly correct for latitude 50° and a flat horizon.

FIGURE 1.6.

crosses the moon's orbit, i.e., when the sun is at a node. (It takes the sun
173.3 days to go from one node to the other.) Because eclipses cannot
occur unless the sun is close to the moon's orbit, eclipses occur only when
the angle between the orbits is close to its maximum.

Why should the angle between the orbits be greatest when the sun is at
a node? Today we know the answer: the change in the angle is caused by
the gravitational pull of the sun on the moon, but the details of the
explanation are complicated.

The Irregular Moon

We have seen that the sun does not move uniformly round its orbit. Nor
does the moon; in fact, it is substantially more irregular than the sun. Its
maximum speed in its orbit is 25% greater than its minimum speed. (The
sun's maximum speed is only 7% greater than its minimum speed.)

Eclipses

Eclipses are striking and occur fairly regularly. (Earthquakes and volcanic
eruptions are also striking, but occur irregularly.) At least, total eclipses
are striking. But once people's attention has been drawn to eclipses and
they look at the moon persistently enough to see the partial eclipses too,
then they will be able to notice some regularity; the more eclipses they
observe, the clearer the regularity will be.

The regularity shows if we count the time-interval between eclipses of the moon. At the time of an eclipse of the moon the earth must be on the straight line joining the sun to the moon, and consequently the moon must be full. Therefore the interval between two eclipses must be a whole number of months. Two successive eclipses of the moon are rarely less than six months apart, and never less than five months apart. If there is an eclipse today, and the next one does not take place in five months' or six months' time, then it cannot take place in less than eleven months' time. In fact, the possible intervals, in months, between successive eclipses are

5   6   11   12   17   18   23   24   29   etc.

Each of these numbers is either a multiple of 6, or 5 more than a multiple of 6. There is a reason for this, too. If the moon moved precisely along the ecliptic it would be eclipsed every month. It does not, however, as remarked earlier (page 13): it moves from about 5° north of the ecliptic to about 5° south, and the time for a complete cycle, say from furthest north to furthest south and back again, is about 27.2 days. This period is called a *latitudinal period*. The time for the moon to go from the ecliptic back to the ecliptic is, of course, half a latitudinal period. The apparent diameter of the moon is about $\frac{1}{2}$°. The earth's diameter is just under four times the size of the moon's. Because the sun is so far away its rays are practically parallel, which means that the earth's shadow is about the same size as the earth, and so, if there were a screen in the sky at the point occupied by the moon, the earth's shadow on it would be a circle of apparent diameter 2°. This means that if the moon is $\frac{3}{4}$° or less from the ecliptic at the time of opposition it will be totally eclipsed; if between $\frac{3}{4}$° and $\frac{5}{4}$° it will be partially eclipsed; if more than $\frac{5}{4}$° it will not be eclipsed. See Figure 1.7.

Let us suppose that at zero time the moon is full and is also on the ecliptic, and so is well and truly eclipsed. One month (about $29\frac{1}{2}$ days) later it will be full again. But 27.2 days after zero time the moon will have returned to the ecliptic, so for just over two days it will have been

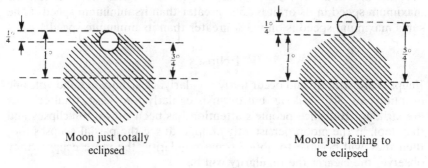

Moon just totally   Moon just failing to
    eclipsed   be eclipsed

FIGURE 1.7.

moving away from the ecliptic. In fact, it will be more than 2° away from the ecliptic and there will be no eclipse. A month later it will be even further from the ecliptic, and so on. But let us see what happens after five months, which is about $147\frac{1}{2}$ days. $5\frac{1}{2}$ latitudinal periods amount to 149.6 days. Thus the moon is two days short of reaching the ecliptic, and so will be 2° away, which means no eclipse. But $29\frac{1}{2}$ is only an average value for the month: months vary because the speeds of the sun and moon vary. If the five months were longer than average they might come close enough to 149.6 days for the moon to be eclipsed, especially if the moon were not exactly on the ecliptic at zero time, but had moved a little way away. Thus five months after an eclipse there could just possibly be another, but there could not possibly be one before then. Six months (177 days) after zero time there is a good chance of an eclipse, because $6\frac{1}{2}$ latitudinal periods is 176.7 days, so the moon is likely to be back close to the ecliptic.

It is because six months is very nearly a whole number of latitudinal half-periods that there is a good chance of an eclipse six months after zero time. We can find other intervals with this property. Here are some examples. On average:

6 months =   13 latitudinal half-periods. Difference: 0.30 days
47 months = 102 latitudinal half-periods. Difference: 0.11 days
88 months = 191 latitudinal half-periods. Difference: 0.07 days
135 months = 293 latitudinal half-periods. Difference: 0.04 days
223 months = 484 latitudinal half-periods. Difference: 0.04 days

Any of these periods could be used for predicting eclipses. The Chinese used the 135-month period (see page 85). The Babylonians used the 223-month period (see page 75). This period eventually became by far the best-known eclipse period and in quite modern times was (mistakenly) given the name "saros." ("Saros" is the Greek form of *saru*, the Babylonian word for 3,600.)

You might wonder whether the same periods could be used for predicting eclipses of the sun. The answer is *no*. The fact that the moon's shadow covers only a very small part of the earth makes delicate calculations necessary; averages are too crude to give correct results. Not until astronomers knew a good deal of detail about the irregularities of the sun's and moon's motions could they predict eclipses of the sun.

The Luni–Solar Calendar

Many early people used the moon for their calendar. Each month started on the day of a new moon and lasted until the day of the next new moon. There were no "30-days-hath-September" complications for these people.

It is natural to fit months and years together as far as possible. Twelve months are just less than a year, so Egyptians, Babylonians, and Greeks had cycles of twelve names for the months. The Chinese simply numbered the months.

The Babylonian names were:

| Nisannu | ayaru | simānu | dūzu | ābu | ulūlu |
| tashritu | arahsamma | kislimu | tebētu | shabatu | adaru |

A civil year started on the first day of *Nisannu*. The next civil year would start twelve months later, which would be about eleven days too early according to the sun. The next year would be twice as much too early, until eventually the civil year is about a whole month too early. To get back into step a thirteenth month was inserted, either a second *ulūlu* or a second *adaru*. At first, the extra months were inserted by royal decree, but from about 528 B.C. they were inserted regularly in an 8-year cycle and from 499 B.C. in a very accurate 19-year cycle [13]:

12 13 12 12 13 12 13 12 12 13 12 12 13 12 12 13 12 13 12

In the 19 years, which total 6,940 days, there are 235 months.

The Greeks used a similar system, the months having different names in different parts of Greece; the cycle was known as the Metonic cycle.

The early Egyptians used the following rule: when Sirius rises in the last eleven days of its month, a thirteenth month is inserted. But about 3000 B.C. they gave up trying to fit the moon to the sun. They divided their year into twelve 30-day months unconnected with the moon, plus five extra days. The Greek forms of later names of the Egyptian months are:

| Thoth | phaophi | athyr | choiak | tybi | mechir |
| phamenoth | pharmuti | pachon | payni | epiphi | mesore |

This calendar got out of step with the sun by a quarter of a day each year, so that if *thoth* is a summer month at one date it will not be a summer month 500 years later.

The Romans also washed their hands of the moon. They started the year on March 1 and made the months alternately 31 and 30 days, cutting short the last month (February) to give the right total. They named the fifth month "July" after Julius Caesar. Later they renamed the sixth month "August" after Augustus Caesar, and changed the 30/31 alternation from then on, so that Augustus's month was as long as Julius's. They also changed the start of the year to January (which is why September is the ninth month although *septem* means *seven*). This is the calendar we have inherited.

The Planets

We mentioned earlier a few starlike objects called planets that move against the background of the constellations. There is a reddish planet

that the Babylonians called *Salbatani*, the Chinese called *Huo xing*, the
Greeks called *Ares*, and we call *Mars*; a rather bright one that the
Babylonians called *Nibiru-Marduk* or *Udaltar* or *Mul-babbar*, the Chinese
called *Mu xing*, the Greeks called *Zeus*, and we call *Jupiter*; and a fainter
one that the Babylonians called *Genna*, the Chinese called *Tu xing*, the
Greeks called *Kronos*, and we call *Saturn*. These planets all move near
the ecliptic: Jupiter, for example, never strays more than $1\frac{1}{2}°$ from it.
Their movement along the ecliptic is quite complicated. Each moves
slowly eastward relative to the fixed stars (i.e., in the same direction as
the sun), gradually slows down, stops, moves westward for a while, slows
and stops again, and finally resumes its eastward motion, repeating this
cycle over and over again. The middle of the retrograde (westward)
motion always occurs when the planet is in opposition to the sun. The
whole cycle takes a more-or-less constant time called the synodic period
of the planet. To take some recent figures as examples, Jupiter started
retrograde motion on February 20, 1970; it started retrograde motion
again on March 23, 1971, giving a synodic period of 396 days. The next
synodic period was 399 days, and the next after that 401 days. The
retrograde motion occupied 124 days of the first cycle, 124 of the second,
122 of the third, and 120 of the fourth. Saturn has a synodic period of
between 379 and 380 days. Mars has a synodic period of around 760 days,
and varies somewhat more than Jupiter: two successive retrogressions (in
1971 and 1973) lasted 50 and 69 days, respectively.

Early astronomers also noticed a bright planet that appeared in the
evening—the Greeks called it *Hesperus*—and an equally bright one that
appeared in the morning—they called it *Eosphoros*. These are, in fact,
the same planet. Homer did not know this, but by 500 B.C. or so the
Greek astronomers did, and called it *Aphrodite*. The early Babylonians
called it *Nindaranna*, the usual Babylonian name for it was *Dilbat* or *Dil-
i-pat*, the Chinese called it *Jin xing*, and we call it *Venus*. It moves close
to the ecliptic and is never more than 47° from the sun along the ecliptic.
Like Mars, Jupiter, and Saturn it retrogresses, but in the middle of its
retrogression it is in conjunction with the sun (not opposition). It cannot,
of course, be seen then.

Finally, there is a fainter planet—the Babylonians called it *gu-ad*, or
gu-utu, the Chinese called it *Shui xing*, the Greeks called it *Stilbon* or
Hermes, and we call it *Mercury*—even nearer the sun. Its synodic period
is between 110 and 125 days.

Venus Observed

For Venus (and theoretically for Mercury, except that Mercury is hard to
see) stargazers can observe more interesting things than just the beginning and end of retrogression. Let us follow Venus through a complete
cycle, starting (i) when it just becomes visible in the morning: it rises just

before the sun and is swamped by sunlight almost at once. Both Venus
and the sun are moving along the ecliptic, the sun eastward and Venus
westward. As Venus gets further from the sun it rises earlier and remains
visible longer until (ii) it ends its retrogression. It starts moving eastward,
slowly at first, and as long as it is moving more slowly than the sun they
are still getting further apart. Eventually, it is moving along the ecliptic as
fast as the sun; at that time it is (iii) at its greatest angular distance west
of the sun. Now it begins to catch up to the sun and when (iv) it is about
10° from the sun it is too near the sun to be seen: it is swamped by
sunlight before it rises above the horizon. Now it invisibly passes the sun
and rises and sets after the sun. As soon as it is about 10° east of the sun
in the ecliptic it (v) becomes visible in the evening, there being time after
the sun's light has dimmed for Venus to be seen before it sets. It continues to move eastward in the ecliptic faster than the sun, and so
increases its angular distance from the sun and the length of time it is
visible, but its speed along the ecliptic is falling. When it has fallen to the
speed of the sun, Venus is (vi) at its greatest angular distance east of the
sun. Later yet, its speed drops to zero and it starts (vii) its retrogression.
This means that it comes rapidly closer to the sun, and when it is about
10° from the sun it becomes (viii) too close to be seen. While it is invisible
it passes the sun, and when it is about 10° behind the sun it becomes
visible in the morning and the cycle starts afresh. The times and positions
(on the ecliptic) of the eight phenomena listed above can be observed and
recorded.

The whole sequence takes from about 575 days to about 590 days, with
average intervals as follows:

End of retrogression (ii)
50 days
Greatest western elongation (iii)
220 days
(Superior) conjunction
220 days
Greatest eastern elongation (vi)
50 days
Start of retrogression (vii)
22 days
(Inferior) conjunction
22 days
End of retrogression (ii again).

The times of first and last appearances ((i), (iv), (v), and (viii)) are quite
irregular, depending, as they do, on the angle between the ecliptic and

the horizon, the amount of haze in the atmosphere, and the keenness of the observer's eyesight. The time between the last appearance in the evening (viii) and the first appearance in the morning (i) varies from two days to twenty days.

For Mars, Jupiter, and Saturn, the times when they become visible in the morning and the times when they cease to be visible in the evening can be recorded. So can the times when they are in opposition to the sun (because they rise as the sun sets and vice versa), and the times when they start and end their retrogressions.

Rising Azimuths

Because the planets move close to the ecliptic they rise and set at positions on the horizon close to where the sun rises and sets. There is some evidence that the Maya were interested in a rising-position of Venus—to be precise, in its most southerly rising-position.

The main evidence is at Uxmal, in Yucatan (latitude about 20° north). A long rectangular building called the Casa del Gobernador, situated on a high mound, is oriented somewhat differently from the other buildings. Anyone looking directly out from the building (i.e., at right angles to the façade) will be looking in a direction 28° south of east, in which direction there is a 25-meter high stone pyramid on the horizon (at Nohpat, several kilometers from Uxmal), now in ruins. This direction corresponds to a declination of $26\frac{1}{2}°$ south; it is 3° too far south for the most southerly rising-point of the sun (3° is six times the diameter of the sun; at this latitude one degree of declination corresponds to one degree of azimuth) and not far enough (by 2°) for the moon, but just right for Venus. Moreover, the Casa del Gobernador is covered with glyphs, among which the glyph for Venus occurs about 350 times. Why the southernmost rising-point should be of particular interest, no one knows [11].

How long would the Maya have to watch, noting the various rising-positions of Venus, until they could be sure that they had found the most southerly one? Not terribly long: eight years. The reason is that, by a remarkable coincidence, five synodic periods of Venus equal eight years almost exactly (and the Maya were well aware of this fact, as is shown by the tables in the Dresden Codex: see pages 199 to 200). This means that every eight years everything about Venus repeats—at least, everything that depends only on the positions of the earth and Venus in their orbits, and that includes the rising-azimuth as seen from any particular place. (Notice that the situation is rather different from that of the moon, which moves round the earth, not the sun, and whose rising-azimuths are affected by the motion of its nodes (see page 13). The nodes of Venus do not move appreciably.)

The Stars

Heliacal Risings

Observations of the rising and setting of stars can be used for a calendar.
We have seen that the sun and the celestial sphere do not turn at quite
the same speed: the celestial sphere is faster by one revolution a year.
Suppose, then, that a star rises at dawn one day. The next day it will rise
a little earlier, the day after that a little earlier yet, until in a year it has
gained a whole day and is again rising at dawn. By watching its rising, we
can tell when a year has elapsed. Actually, a star rising exactly at dawn
cannot be seen—the sky is too bright. (The sky lightens before dawn and
stays light a little after sunset because of refraction.) Thus instead of the
true heliacal rising (rising simultaneously with the sun) astronomers had
to use the first visible heliacal rising. On the day before the first visible
rising the star rose so close to the sun that it could not be seen. On
subsequent days it rose earlier and earlier, and was visible in the eastern
morning sky for longer and longer before being swamped by sunlight.

Both the ancient Egyptians and the early Greeks used heliacal risings
as a calendar: Hesiod's *Works and Days* (about 800 B.C.) is a kind of
farmers' calendar in verse and makes recommendations like "pick the
grapes when rose-fingered Dawn looks on Arcturus."

Precession

This is the first irregularity mentioned on page 7—the fact that the center
of rotation of the celestial sphere, the celestial pole, is not fixed.

Nowadays, the north celestial pole is close to a fairly bright star—the
Pole Star. However, as the centuries pass the pole moves and traces out a
circle on the celestial sphere, taking 26,000 years for a complete circuit.
There are not many bright stars on this circle, and there was no pole star
in Babylonian, ancient Greek, or ancient Chinese astronomy. The ecliptic
is fixed on the celestial sphere, so this movement means that the equator
moves relative to the ecliptic. In fact, it moves in such a way that the
angle between the two does not change.

If we consider motion relative to the sun, we can visualize precession as
follows. Imagine a model of the solar system set up on a table-top, with
the earth moving round the sun, the table-top representing the plane of
the ecliptic. The earth's axis will not be vertical, but will make an angle of
$23\frac{1}{2}°$ to the vertical, the same as the angle between the equator and the
ecliptic (i.e., the table top). In a short period of time, a few years say, the
direction of the earth's axis will not appear to change, but will point in
the same direction in space no matter where the earth is in its orbit. But
in the long run the axis will rotate, still keeping the same angle from the
vertical. In fact, the earth's axis is behaving rather like the axis of a

spinning top, and its rotation can be explained equally well by the laws of motion in modern physics.

The above description is not absolutely accurate. The angle between the equator and the ecliptic is not quite fixed; it has changed (see page 9) by about one-fifth of a degree in 5,000 years. But precession covers about 72° in 5,000 years (360° in 26,000 years), compared with which the change in the angle of tilt is negligible. The path of the pole on the celestial sphere is not quite a circle, but it is close enough for all practical purposes.

Precession has two important consequences. First, because the pole moves relative to the celestial sphere, the distance of a star from the pole changes as time passes. Consequently, its directions of rising and setting change, and investigations of the alignment of ancient edifices to the risings and settings of stars must allow for this.

Second, at a summer solstice the sun will be at a certain position on the ecliptic. In a year, the sun will move round the ecliptic back to the solstice. But during this year the solstice position will have moved relative to the stars, and the sun will not yet be back in the same position relative to the stars: that will take another twenty minutes or so. Thus there are two different years. The time from solstice to solstice (or equinox to equinox) is the *tropical* year; and the time for a complete revolution relative to the stars is a *sidereal* year. The tropical year is the important one for practical purposes, and the word "year" on its own means tropical year.

Because the orbit of the sun on the celestial sphere is not quite fixed the description on page 9 was a little oversimplified.

Precession and the Pyramids

Precession had an effect on the orientation of the pyramids, which were aligned with remarkable accuracy. The eastern and western sides of the "Great" Pyramid (the pyramid of Khufu) run almost exactly due north and south: in fact, they point only 330 and 150 seconds west of north. (330 seconds is less than one-tenth of a degree.) The other two sides are even more accurate: they point 148 and 117 seconds south of west. The errors here are all clockwise. We have the following table for the errors in orientation of third- and fourth-dynasty pyramids. (A negative error is counterclockwise.) The dates listed are the beginnings of the reigns of the pharaohs to whom the pyramids are attributed.

Pyramid at Maidum	0.5°	2637 B.C.
"Bent" pyramid at Dahshur	0.2°	2613 B.C.
Pyramid of Khufu at Giza	less than 0.1°	2589 B.C.
Pyramid of Khafre at Giza	0.1°	2558 B.C.
Pyramid of Menkaure at Giza	−0.2°	2530 B.C.

Clearly the orientation is turning steadily counterclockwise as the years pass, though Khafre's pyramid does not quite fit the pattern. Is this a coincidence, or is there a reason for it?

Let us consider how the Egyptians could have oriented the pyramids. They must have started by finding north or south—east and west are simply the directions midway between them and cannot be found directly. To find north or south accurately we must use the stars; the sun is too big and too bright to yield an accurate result. I describe one way in which north can be found on page 26. Another way would be to bisect the angle between the directions in which a star rises and sets, taking care that the horizon heights are the same at the two points, perhaps by building an accurately leveled wall some distance north of the observer and using this as a horizon. All these methods are long and complicated, but once north has been found at some place, all four directions are known. Having found east and west we can wait until we see a star that rises due east or sets due west. Now if we want to find the cardinal directions at some other place we need not go through the long process of finding north again; we simply watch for our star. If the Egyptians used this method to orient the pyramids, this means that for each pyramid they established the east–west alignment first and obtained the north–south alignment from it. This agrees with the fact that the east–west directions in Khufu's pyramid are more accurate than the north–south directions. More important—as the years pass, precession will cause the direction in which our star rises or sets to change, and so the orientation will change, as shown in the table. In fact, there is a star, namely β Scorpii, whose rising-directions at the various dates agree precisely with the orientation of all the pyramids listed above, except Khafre's; this pyramid is aligned on the direction in which β Scorpii sets [110].

The Astronomer's Tools

Let us see how to use some instruments that early astronomers might have used. Using a plumb-line, erect a vertical rod or pillar on a flat stone base leveled by means of water channels carved in it. The instant when the shadow is shortest is noon. The shadow always points in the same direction at noon; this direction is north (in the northern hemisphere). For greater accuracy, use a star instead of the sun. Knowing roughly where south is, place yourself to the south of the rod (or of a plumb-line) and watch a bright star as it moves round the celestial pole. Keep yourself in line with the star and the rod, and when the star is just at its highest elevation and beginning to descend drive a peg into the ground. The line from the peg to the rod will run pretty accurately north and south. Once the direction of south has been found, it is much easier (and more accurate) to determine noon by noting when the sun is due south rather than by noting when shadows are shortest.

FIGURE 1.8. Chinese instrument (*gui biao*) for measuring shadows at noon.

The Chinese used vertical rods extensively for finding the height of the sun at noon. They would lay a stone ruler along the ground running due north from the foot of the rod. At midday, the shadow of the rod would lie along the ruler and could be measured. Sometimes the ruler would be a permanent structure, with the rod fitted into a socket at its south end. For most of Chinese history, the length of the rod was standardized at 8 *chi* (just under 2 meters). See Figure 1.8.

Ptolemy (about A.D. 150) described an ingenious device for finding the elevation of the sun at noon. A block of stone is set up so that one face is vertical and in the north–south direction (see Figure 1.9). A small peg is fixed at the middle of an angular scale engraved on it. The shadow of the peg falls across the scale and disappears at noon when the whole face is shadowed. The astronomer takes the scale reading at this instant [14].

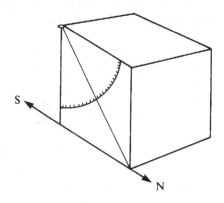

FIGURE 1.9.

FIGURE 1.10.

As soon as people became adept at metal working they could make thin straight sighting-tubes or sighting-rods. The Chinese preferred tubes, perhaps because they had earlier used bamboo tubes; Europeans preferred open sights.

A tube pivoted at the center of a quadrant which pivots about a vertical axis could be used to measure the elevation of a star; a horizontal scale would give its azimuth (see Figure 1.10). The elevation of a star when it crosses the vertical north–south plane is especially useful. To find this we can use a scale marked on a north–south wall, which can be firmer and more accurate than a movable quadrant (Figure 1.11).

Instruments with circular scales like those in Figures 1.9–1.11 have one particular advantage over a vertical rod for measuring the elevation of the sun. This is because the sun is not a point, but has an appreciable size: its apparent diameter as seen from the earth is about $\frac{1}{2}°$.

In Figure 1.12 (which is far from being to scale) the sun casts a shadow of the rod PQ. The stretch QR is in deep shadow (assuming that the rod is thick enough to block the view of the sun, i.e., that the apparent width of the top as seen from R is not less than $\frac{1}{2}°$. This means, if the rod is about 2 meters high, that it is more than about 3 cm thick). The stretch RS is lit by part of the sun's disk, and the shadow grows progressively lighter from R to S, where it disappears altogether. There is no way in which the eye can judge the point T, about halfway between R and S,

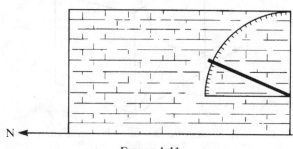

FIGURE 1.11.

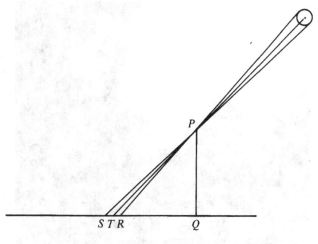

FIGURE 1.12.

that corresponds to the center of the sun. In fact, the eye considers the shadow as ending at (or very close to) R, where the shadow begins to fade. Thus the measured height is apt to be too high by half the sun's apparent diameter, i.e., about $\frac{1}{4}°$.

The Chinese, however, eventually developed the vertical rod into a very accurate instrument and, in particular, they overcame the disadvantage we have just mentioned. The Yuan dynasty astronomer Guo Shoujing (A.D. 1231–1316) erected a rod 40 chi high, five times the standard height. The rod was 50 chi long, with 15 chi buried in the ground for stability, so that the top was 35 chi above the ground; the other 5 chi was accounted for by two bronze dragons holding between them a horizontal crossbar aligned in the east–west direction. The scale for measuring the shadow was about 30 meters long, and accurately leveled by means of two water-filled grooves running its whole length. Because of the distance of the crossbar from the scale, its shadow will be spread out and faint: there will be no full shadow (umbra), only partial shadow (penumbra). To overcome this, Guo used a device called a ying fu (shadow definer). This consists of a small sheet of copper pierced by a small hole about 2 mm across. It is set in a stand which can be moved along the scale, and which allows the sheet to be swiveled until it is perpendicular to the sun's rays. The sun shining through the hole produces a spot of light in the middle of the shadow of the copper sheet. When the hole is in line with the crossbar and the sun, a small sharp image of the crossbar is seen running across the spot of light. If the hole is slightly out of position the image of the bar will cut the spot of light into two unequal parts; the astronomer has to adjust the position of the ying fu until the image of the bar cuts the spot of light exactly in half. Its

FIGURE 1.13. Tang dynasty gnomon. At midsummer noon the shadow just reaches the edge of the base. (Photograph Robin Rector Krupp, Griffith Observatory.)

position on the scale then gives a precise reading for the position of the center of the shadow of the crossbar. In effect, the *ying fu*, by blocking all the sunlight except the narrow beam passing through the hole, converts the sun from a half-degree-wide source of light to almost a point-source.

At Dengfeng, in Honan Province, there is an impressive building, attributed to Guo [16]. It consists of a brick tower about 9 meters high, the top being a square platform about 8 meters by 8 meters (plenty of room for astronomers to work), and the faces sloping so that the base is substantially bigger than the top, see Figures 1.14 and 1.15. The edges of the platform are oriented in the north–south and east–west directions. From the middle of the north edge a vertical channel is cut in the north face of the tower; it is quite deep at the bottom because of the outward slope of the face, and from the bottom of this channel, which

FIGURE 1.14. Tower at Dengfeng seen from the end of the horizontal scale. (Photograph Robin Rector Krupp, Griffith Observatory.)

is vertically below the edge of the platform, runs a shadow-measuring scale of the sort just described. If a horizontal crossbar were placed 40 *chi* above the end of the scale it would be about waist level for someone working on the platform, a convenient height. A plumb-line from the crossbar down the channel would enable the astronomers to find the zero point of the scale quite accurately.

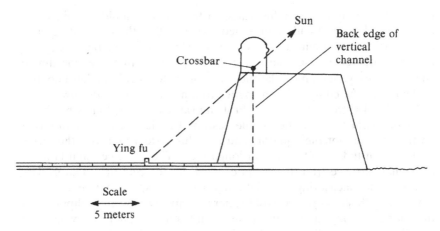

FIGURE 1.15. Tower at Denfeng: elevation.

Islamic astronomers were also well aware that the bigger the instrument the more precise the observations. The eleventh-century astronomer Ibn Qaraqa went as far as to tell the vizier who was financing his observatory that if he were given enough funds to make an instrument that stretched from the pyramids to a mosque on Mount Muqattam he would certainly do so [16a].

The tendency to increase the size of instruments is well exemplified by the observatory of Ulugh Beg in Samarkand, begun in A.D. 1424. It fell into ruins shortly after his death and was not rediscovered until 1908, but enough remains to suggest what it was probably like [16b]. It consisted of a circular building of over 50 meters diameter and about 35 meters high. The main instrument was a quadrant of 40.4 meters radius in the vertical north–south plane, part of the ground being excavated to accommodate it: see Figure 1.16(a). The quadrant was double, being constructed of two parallel walls (see Figures 1.16(b) and 1.17). They were just far enough apart for an astronomer to sit between them and make observations, and were marked in degrees by grooves running across them. The back-sight was probably a wooden rectangle equipped with handles by means of which two assistants on the outer steps could set it down accurately between two degree markings; possibly two ridges on the underside, the right distance apart, fitted into two successive degree grooves: see Figure 1.16(c). In the middle of the rectangle was a slot along which the observer could move a peep-hole and read off the fractions of a degree on a scale alongside the slot. The front sight was an opening at the top corner of the building and would be furnished with cross-wires marking the center of the circle of which the quadrant formed part. The vertical cross-wire would then mark due south.

Celestial Latitude and Longitude

Before describing more sophisticated instruments I should explain these two terms. I have already mentioned (page 6) that the celestial equator can be used to pinpoint a position in the sky in the way that the geographical equator is used on earth, and (page 7) that the coordinate analogous to geographical latitude is not called celestial latitude but declination. The coordinate analogous to longitude is *right ascension*.

The Greeks, from the time of Ptolemy (at least) and, following them, astronomers in Europe, the middle east, and India, used a system based on the ecliptic, not the equator, and it is this system that uses the terms "celestial latitude" and "celestial longitude." Therefore celestial latitude is angular distance from the ecliptic, not from the celestial equator. The zero point for measuring celestial longitude is the spring equinox.

Figure 1.18 shows the celestial sphere, centre O, and the ecliptic. A is the spring equinox, and P and Q are points on the sphere for which the line POQ is perpendicular to the plane of the ecliptic. X is a point whose

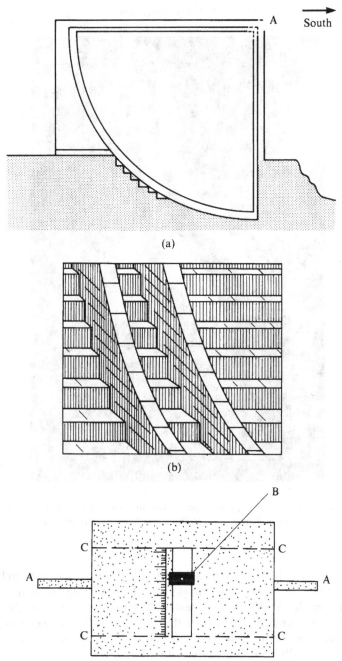

(a)

(b)

B

C   C

A   A

C   C

AA: handles.
B: the peep-hole that can be moved up and down the slot.
CC: ridges one degree apart on the underside.

(c)

FIGURE 1.16.

FIGURE 1.17. The remaining portion of the meridian arc in Ulugh Beg's observatory at Samarkand. (Photograph by Ernest W. Piini.)

position we are interested in. X^* is the point where the semicircle PXQ
crosses the ecliptic. Then the angle X^*OX is the celestial latitude of X,
and the angle AOX^* is its celestial longitude. (To us, longitude and
latitude are technical terms, but the Greek words that we translate by
longitude and latitude are the ordinary Greek words for length and
breadth, i.e., distance along and distance across.)

It is interesting to note the effect of precession on the latitude and
longitude of a star. The latitude is the angular distance from the plane of
the ecliptic, and this does not change as the earth's axis rotates. The
longitude is the angular distance round the ecliptic, measured from one of
the points where the equator crosses the ecliptic, and this point does

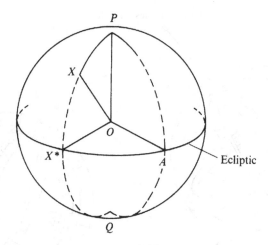

FIGURE 1.18.

move as the axis rotates. Consequently, as time passes, latitudes remain the same but longitudes increase at a steady rate of 72° per century.

Armillaries

If a movable quadrant were set up with its axis pointing to the pole instead of being vertical, as it is in Figure 1.10, it would measure declination and right ascension. We know of no quadrants set up like this, but if we complete the quadrant to a whole circle we get a simple form of armillary sphere. (*Armilla* is the Latin for "ring.") Figure 1.19(a) shows the principle: QP points toward the pole and is the axis about which the ring B pivots; the ring C is at right angles to PQ. The whole is usually supported by an outer ring A, to which C is rigidly fixed, and to which B is pivoted. The Chinese used such armillary spheres extensively [17]. We do not know for certain whether the early Greeks used them, but Ptolemy described a development with more rings and an extra pivot which could be used to measure celestial latitude and longitude directly.

Ptolemy called the instrument mentioned above an *astrolabon* [18]. It is not like the well-known medieval astrolabe, even though the names are so alike. To make an astrolabon, start with an armillary sphere as shown in Figure 1.19(a). Figure 1.19(b) shows it without its base. Fit it with pivots H and K placed so that the angle between HK and PQ is equal to the angle between the ecliptic and the celestial equator. Graduate ring C (in degrees, or however you want to measure longitudes). Graduate ring B (Figure 1.20(a,b)) in degrees (or however you want to measure latitudes). Outside ring A pivot a ring D so that it is free to turn about the same axis as B. Pivot the ring A inside a fixed framework, as shown in

To north celestial pole

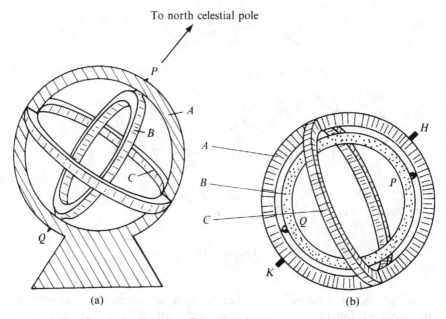

FIGURE 1.19.

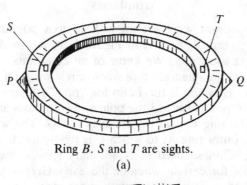

Ring *B*. *S* and *T* are sights.

(a)

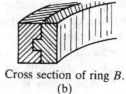

Cross section of ring *B*.

(b)

FIGURE 1.20.

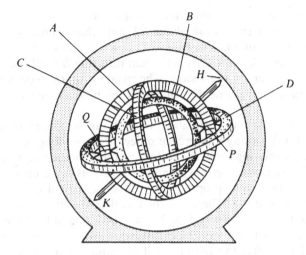

FIGURE 1.21.

Figure 1.21, setting it up so that *HK* points to the celestial pole. Then, because the angle between *HK* and *PQ* is the angle between the ecliptic and the equator, it is possible for ring *C* to swing into the plane of the ecliptic as rings *A* and *C* pivot round *HK*.

FIGURE 1.22. Guo Shoujing's armillary (see Figure 1.26). (Photograph Robin Rector Krupp, Griffith Observatory.)

The astrolabon is used when the sun and the moon are both visible. First turn the inner framework about HK until the shadow of one half of ring C falls on the other half. Ring C is now parallel to the ecliptic, and the $A-C$ framework is properly set. Then turn ring B until the sights line up with the center of the moon. The reading of the sights gives the moon's latitude, and the reading of ring B on the scale on ring C gives its longitude.

We can also use the astrolabon to find the longitude of a star or planet—call it X—using a reference object Y whose longitude is known. To do this, set the ring D on the scale C to the longitude of Y. Then turn the movable framework as a whole until ring D is lined up with Y. The $A-C$ framework is now properly set, and we can read the latitude and longitude of X as above. Therefore the Greeks, who had tables giving the longitude of the sun, could find the latitude and longitude of the moon by using the sun as a reference object instead of using the shadow method. They could then find the latitude and longitude of a star by using the moon as a reference object. The procedure is as follows:

(1) Just before sunset, find the longitude of the moon as described above, calculate the change in longitude of the moon in half an hour (allowing for the change in parallax), and set ring D to the new longitude.

(2) When the half-hour has elapsed and the stars are visible, line ring D up on the moon and ring B on the star whose latitude and longitude are wanted.

(In an example given by Ptolemy, however, the correction for the change in the moon's longitude and parallax was made afterward, and in the second step ring D was set to the longitude found in the first step. Moreover, in the first step, ring D was set to the sun's longitude calculated only to the nearest degree and the correction for the odd fraction of a degree was made later.)

After the longitude of one star has been found, we can find the latitude and longitude of as many stars as we like by using this one as a reference object.

The Chinese presumably used armillaries about 350 B.C., as this is the date of two early star catalogues. One armillary, built by Kong Tong in A.D. 323, was like the one shown in Figure 1.19(a), except that the outer framework consisted of three rings welded together, the third ring being horizontal; and the rotating ring was a split ring, carrying a sighting-tube between its two parts. The vertical ring was also split, presumably to allow a sighting-tube to be used on the meridian.

Later, other rings were added, including an ecliptic ring and a ring in the plane of the moon's orbit, whose points of attachment could be moved as the nodes of the moon regress. Later yet, when astronomers

FIGURE 1.23. The pointer on the equatorial ring of Guo Shoujing's armillary.
(Photograph Robin Rector Krupp, Griffith Observatory.)

FIGURE 1.24. Armillary now at Nanjing, probably a replica of one at Beijing
described by Guo Shoujing. (Photograph Robin Rector Krupp, Griffith Observatory.)

FIGURE 1.25. Model of Su Sung's clockwork-driven armillary. (Photograph E.C. Krupp, Griffith Observatory.)

found that for extreme accuracy a firm stable apparatus was essential, armillaries were simplified, and one built by Guo Shoujing in 1276 had a single movable ring carrying a sighting-tube and rotating about an axis pointing to the pole, see Figures 1.22 and 1.26. The fixed ring used for measuring the position of the rotating ring was at the bottom end of the axis instead of at the center, and there was another, smaller, one at the top end.

The most interesting development, in which the Chinese advanced

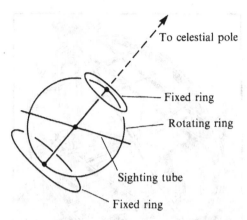

To celestial pole

Fixed ring

Rotating ring

Sighting tube

Fixed ring

FIGURE 1.26. Guo Shoujing's armillary.

beyond anything available in the West until the observatory telescope of modern times, was the use of a power-drive (a water-wheel), see Figure 1.25. The first mechanical armillary was described about A.D. 125. Later, after the invention of clockwork in the eighth century, armillaries could be controlled to keep pace with the heavens, just like today's telescopes [19]. At the end of the sixteenth century, instruments in the observatory at Nanjing were described by an Italian visitor, Matteo Ricci, as "all cast of bronze, carefully worked and gallantly ornamented, and so large and elegant that none better had been seen in Europe," see Figure 1.24.

In the west, too, large simplified instruments were developed. One example is Tycho Brahe's *armillae equatoriae maximae* of around A.D. 1580, shown in Figure 1.27.

Polar Elevations and the Obliquity of the Ecliptic

Besides the quantities that can be read directly from these instruments there are others that can be deduced quite simply.

Although the elevation of the sun can be read directly from an instrument furnished with an angular scale, like Ptolemy's plinth (Figure 1.9) or a wall quadrant (Figure 1.11), anyone using a vertical rod must deduce it from the length of the shadow. The Chinese did this by drawing an accurate scale diagram and measuring the angle: as late as A.D. 725 an account of an expedition to measure shadows mentions a "right-angled triangle diagram," though by the Yuan dynasty (the time of the giant rod and the *ying fu* mentioned above) the Chinese had the equivalent of trigonometric tables.

The elevation of the celestial pole varies from place to place, being equal to the latitude of the place of observation (see Figure 1.28). Even

Armillary from Stjerneborg

FIGURE 1.27.

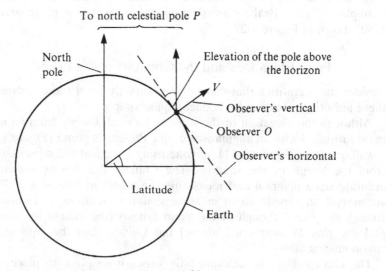

FIGURE 1.28.

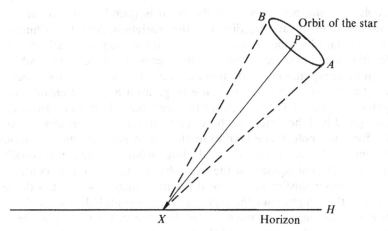

X = observer.   Angle HXP = elevation of the pole.
P = celestial pole.   Angle HXA = least elevation of the star.
                               Angle HXB = greatest elevation of the star.

FIGURE 1.29.

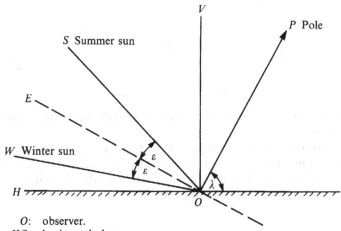

O: observer.
HO: horizontal plane.
OV: vertical.
OP: toward the celestial pole.
OE: at right angles to OP, and so parallel to the celestial equator.
OS: toward the summer solstice noon sun.
OW: toward the winter solstice noon sun.
  λ: the observer's latitude.
  ε: the obliquity of the ecliptic.

FIGURE 1.30.

people who did not know that the earth is round would, if they made widespread observations, discover this variation, and the Chinese recorded polar elevations at various places just as we record latitudes. Polar elevation cannot be measured directly because the pole is not visible, but it can be computed easily by two methods. The first uses a star near the pole (see Figure 1.29). We measure its greatest and least elevations as it circles the pole. Exactly halfway between them is the elevation of the pole. (And half the difference between them is the angular distance of the star from the pole.) The second method is to measure the elevations of the sun at noon at the summer and winter solstices. The sun is equally far from the celestial equator at the two solstices, having a declination of $+\varepsilon$ at the summer solstice and $-\varepsilon$ at the winter solstice, where ε is the angle between the ecliptic and the equator (the obliquity). In Figure 1.30, the observer's latitude λ is shown as the elevation of the pole, and the angle *HOE* between the equator and the horizontal is $90° - \lambda$. We measure the elevation *HOS* of the summer sun: call it α. Then

$$\alpha = 90° - \lambda + \varepsilon.$$

If β is the elevation *HOW* of the winter sun,

$$\beta = 90° - \lambda - \varepsilon.$$

Then

$$\varepsilon = \tfrac{1}{2}(\alpha - \beta)$$

and

$$\lambda = 90° - \tfrac{1}{2}(\alpha + \beta).$$

Thus from the two elevations of the sun we find both the obliquity and the latitude.

This method has the practical difficulty that the solstice will usually occur not at noon but between noons, so the sun will not have reached its extreme declination when the measurement is made. We will see on page 98 how the Chinese dealt with this difficulty.

CHAPTER 2

Megalithic Astronomy

Stonehenge

Everyone knows of the massive group of stones on Salisbury Plain, impressive even as ruins, see Figure 2.1. If we stand in the middle of them at dawn on the day of the summer solstice, we can watch the sun rise over a distant stone pillar known as the heel-stone. The people who built Stonehenge must have been more than just interested in midsummer sunrise; they must have been eager to mark it on a vast scale [21].

Figure 2.2 is a plan of Stonehenge as it is now. The surviving stones have been given numbers, and if a missing stone has left a clear-cut hole, the hole has been given a number or a letter. Before Stonehenge fell into ruins it was probably as shown in Figure 2.3.

The bank that forms the outer perimeter is made of chalk and was originally about 2 meters high. The heel-stone (stone 96) is about 5 meters high and just cuts the skyline when seen from the center of Stonehenge; it is not quite in the center of the avenue formed by the two straight banks that run toward the north-east. The four stones 91–94 are known as the "stations." They form a rectangle whose short sides are parallel to the avenue.

The larger of the stone circles in the middle consisted of thirty uprights, about $4\frac{1}{2}$ meters high, carrying thirty horizontal stones, each bridging the gap between one upright and the next. The material is sarsen, a kind of sandstone. Although the ground on which Stonehenge was built slopes a little, the circle formed by the horizontal stones is thought to have been level; the upright stones on the downhill side are a trifle taller than those on the uphill side. The smaller circle, not much of which is left, is made of bluestone (various types of igneous rock, indigenous to Wales). Inside this circle were five sarsen trilithons, each consisting of two uprights set close together with a third stone across the top; the biggest is more than 7 meters high. The innermost U-shaped blue-stone structure is badly ruined.

FIGURE 2.1. View from the center of Stonehenge, showing the heel-stone just cutting the horizon. (Photograph E.C. Krupp, Griffith Observatory.)

B, C, D, E, F, G, and H are stone-holes. The various post-holes once supported wooden posts. The "car-park" post-holes are about 275 meters from the center and are right on the line 92 → 93. However, these holes have been dated (by applying the radio-carbon method to charcoal found in them) and are two or three thousand years earlier than the rest of Stonehenge. They probably have nothing to do with the later structures [22].

Finally, there are 56 holes in a circle round the inside of the bank. They are known as "Aubrey holes," and have never supported any kind of upright, whether wood or stone. In some of them such things as antlers, bone pins, and the remains of human cremations were buried. Some commentators think that the Stonehenge astronomers used them as a counting device, moving pebbles from one hole to another to keep track of the motions of the sun and moon [23].

Stonehenge was not built all at once. The chalk bank was probably built first, about 3200 b.c., the sarsen circle and trilithons coming perhaps a thousand years later. The bluestones were first erected at some time between these two dates, but were re-erected twice, being placed in their final position after the sarsens were erected. Let us think of Stonehenge in two parts:

Outer Stonehenge: The bank, avenue, stations, post- and stone-holes, heel-stone, and Aubrey holes.

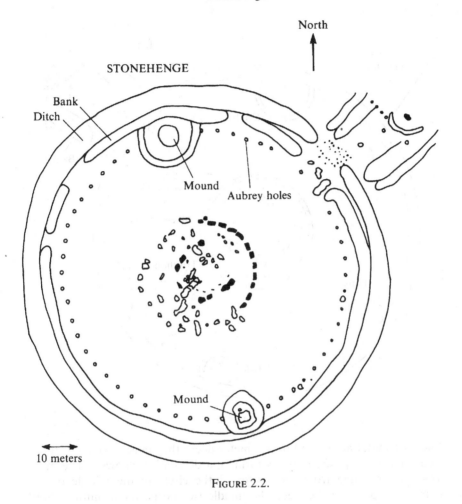

North

STONEHENGE

Bank
Ditch

Mound
Aubrey holes

Mound

10 meters

FIGURE 2.2.

Central Stonehenge: The sarsen circle and trilithons, and the blue-stones in their final positions.

Outer Stonehenge and Central Stonehenge do not have quite the same center. The two centers are about half a meter apart—not far enough to produce much ambiguity in the direction from the center to the heel-stone. (The difference is about 0.1° [24].)

You will notice in Figure 2.2 that the avenue does not fit exactly onto the entrance through the bank and ditch. The reason is that the avenue was constructed much later than the entrance, and the orientation was deliberately altered. The gap in the ditch shows where the entrance originally was. As seen from the center of Stonehenge, the heel-stone was practically on the right of the entrance way. To the left of, and slightly in front of, the heel-stone are four large post-holes, marked A in Figure 2.3.

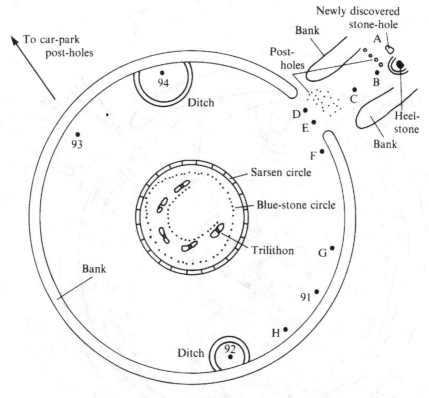

FIGURE 2.3.

Each is a meter across and over a meter deep; they must have held sturdy tree-trunk sized posts, perhaps forming a decorative gateway. The main axis, i.e., the line from the center of the circle to the middle of the entrance gap, passes between the middle two posts. Its azimuth is 46.6° (clockwise from north).

Across the entrance gap itself are 53 small post-holes, labeled simply "post-holes" in Figure 2.3.

Later, probably about the time the sarsens were erected, the avenue was formed by constructing the bank and ditch that outlines it. The right side of the avenue was well to the right of the right-hand edge of the old entrance gap, and presumably part of the ditch was deliberately filled in and the corresponding part of the bank removed, to make the new entrance fit the avenue. The new axis has azimuth 49.9°, and the heel-stone is only slightly to the right of it (its azimuth from the center is 51.1°) [27].

We have seen that the line from the center of Stonehenge to the heel-stone points close to midsummer sunrise. In precisely what direction did the sun rise in 3200 B.C.? Unfortunately, not quite the same direction as now, or we would only have to go and look. This is because the angle between the ecliptic and the equator has changed slightly. The exact answer depends on what we mean by the moment of sunrise. If we take it to be the instant when the top of the sun first appears, we get one answer; if we take the first instant when the whole of the sun's disk is visible we get a different answer, because the sun comes up obliquely. The height of the horizon behind the heel-stone is 0.6° [25]. Assuming that it was much the same in 3200 B.C., the answers are:

49.3° for "first gleam" sunrise,
50.2° for "whole disk" sunrise [26].

Therefore the first gleam of the sun appeared 2° to the left of the heel-stone. When the sun appeared to be standing on the horizon it was still 1° to the left. By the time it was directly over the heel-stone it was a full sun-diameter above the tip of the stone. This may well be what the builders of Stonehenge intended, but there is another possibility. In 1979 a stone-hole was discovered about 3 meters to the left of the heel-stone. (It is not shown in Figure 2.2.) Its azimuth from the center is 48.3°, and if a stone similar to the heel-stone occupied it the pair would have formed a spectacular foresight, framing the sun rising in the gap between them.

The small post-holes in the entrance are in just the right position to mark northerly moonrises. The moon stays close to the ecliptic (the sun's path) and rises and sets close to where the sun does. Consequently, most moonrises, like the sunrises, lie between the northernmost and southernmost sunrise. But not all. At those times when the swing of moonrise is greater than the sun's (see pages 13 to 15), the moon will sometimes—once or perhaps twice in each $27\frac{1}{2}$-day cycle—rise outside the limits set by the sun. The Stonehenge astronomers will have been well aware of these limits, and possibly have regarded them as some kind of mysterious barrier, not normally transgressed. At any rate, they thought moonrises further north than the northernmost sunrise worth marking, and set up posts for this purpose. For half of an 18.6-year cycle all moonrises will lie inside the limits set by the sun; when the moonrise swing has widened enough the moon will begin to rise occasionally further north than the sun ever reaches, and will continue to do so until the swing has shrunk too much for the moon to trespass outside the sun's boundaries. The post-holes are arranged in six rows, so it looks as though the Stonehenge astronomers observed the moon through six of these cycles. They decided (perhaps for some reason connected with a moon-ritual) to make the entrance fit the arc covered by the post-holes. This is, of course, speculation, but no other explanation that I have seen explains why there are post-holes only to the left of the heel-stone and none on the right to mark the directions of northernmost moonrise at those times when the moon-swing is narrower than average. (These would lie between the heel-stone and the direction marked M_2 in Figure 1.6.)

The short sides of the station rectangle are not quite parallel to the axis of Stonehenge. The azimuths are 49.8° and 50.6°, as against 49.9°, for the (new) axis [27].

If a certain direction, say from X to Y, points to midsummer sunrise then automatically the opposite direction, from Y to X, points pretty close to midwinter sunset. Thus the short sides of the station rectangle automatically mark midwinter sunset if we look from 91 to 92 and from 94 to 93. It is true that stone 93 is very short, not much over 1 meter high, but even so it cuts the horizon when seen from 94, and so could be used for sighting [28].

(The reasons why midwinter sunset is not exactly opposite to midsummer sunrise are:

(i)   the effects of refraction;
(ii) the fact that the visible horizon is a little above the horizontal; and
(iii) the fact that we are not taking sunrise to be the instant when the
      center of the sun's disk appears.

Although the direction opposite to first-gleam midsummer sunrise is 229.3°, the direction of last-gleam midwinter sunset is 229.8°.)

The long sides of the station rectangle point to the southernmost moonrise and the northernmost moonset [29]. To avoid complication, let us concentrate on the risings and ignore the settings for the moment. Presumably the builders noticed that the directions of the northernmost sunrise and the southernmost moonrise are at right angles when seen from Salisbury Plain (see Figure 1.6, the latitude of Stonehenge is 51.18°) and incorporated a rectangle in their complex to mark this fact. One of the diagonals, the line $93 \rightarrow 91$, points to the southernmost moonrise when the swing is least. Figure 2.4 shows how all the directions in Figure 1.6 are marked at Stonehenge [30].

This is more sophisticated than the arrangement in the entrance, because finding the southernmost rising when the swing is least requires some knowledge of the cycles of the moon's movements, whereas finding the southernmost (or northernmost) rising of all requires only patience.

Central Stonehenge may also show alignments [31]. An observer wedged between the two uprights of the eastern trilithon can see midwinter sunrise between two of the uprights of the sarsen circle (see Figure 2.5). From the next trilithon, the southernmost moonrises show between uprights. The two trilithons on the other side similarly reveal the northernmost sunset and moonsets. The central trilithon, facing directly opposite the main axis, automatically reveals midwinter sunset. These alignments are not as precise as those for Outer Stonehenge: a gap between uprights is vaguer than a stone to be sighted over. Nor is the set-up complete: Central Stonehenge does not mark the northernmost moonrises (nor southernmost moonsets).

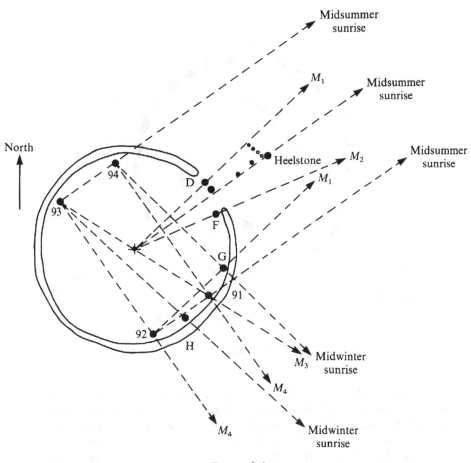

Midsummer sunrise

Midsummer sunrise

Midsummer sunrise

Midwinter sunrise

Midwinter sunrise

FIGURE 2.4.

The fact that the main axis of Stonehenge points to midsummer sunrise has been known for a long time. The first mention that I am aware of is by William Stukely in 1723 [32]. The suggestion that Stonehenge also marked moonrise directions was first made by G. Charrière in 1961, and the full set of directions in Figure 2.4 is due to Gerald Hawkins [30].

There is an interesting structure at Sarmizegetusa in Romania, which, like Central Stonehenge, is of the form of a horse-shoe inside a circle (Figure 2.6). In this case the horse-shoe opens out toward midwinter sunrise. The structure is much smaller and later than Stonehenge, and made of wood, not stone [33].

There are also U-shaped constructions at Dun Ruadh in Ireland and at Loanhead of Daviot and Croft Moraig in Scotland. The one at Croft Moraig, though much smaller than Stonehenge, has some interesting

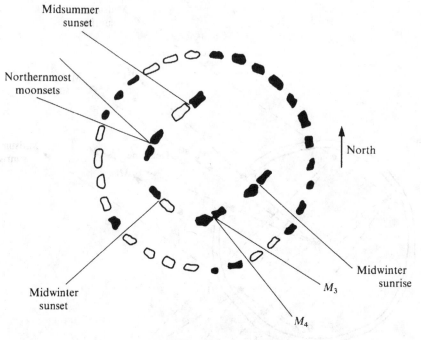

Midsummer
sunset

Northernmost
moonsets

North

Midwinter
sunrise

M_3

Midwinter
sunset

M_4

FIGURE 2.5.

similarities (and differences). It was reconstructed over the years (from
wood to stone), it had stones arranged in the shape of a U (but opening
toward the south-south-east) inside a stone circle which itself was inside
(not outside) a circular bank. It had pillars, presumably marking an
entrance (but at the east, not at the opening of the U—opposite the
opening was a long stone slab). The alignment of the U is too far south to

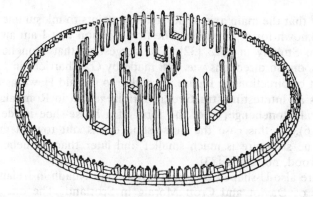

FIGURE 2.6. Reconstruction of the structure at Sarmizegetusa.

mark a sunrise but it does point roughly toward the southernmost moon-rise [34].

How much astronomy could be understood by the people living in the British Isles at the time the megaliths were erected? They had no writing, so the vital question is, "How much astronomy is possible without writing?" Anyone can find rising- and setting-directions of stars, or the northernmost and southernmost rising- and setting-directions of the sun and moon without writing. It is even possible to count the number of days in a year or the number of years between one maximum moonrise-swing and the next without writing; for example, by using notches on a tally-stick or scratches on a bone. In fact, there is evidence that early people kept track of the phases of the moon in this way. However, there is no evidence that any prehistoric people found the exact day of the northernmost sunrise, and although today tourists (informed by diaries and almanacs) throng Stonehenge on the day of the solstice, in megalithic times people may, for all we know, have gathered there for several days before and after the solstice.

We should not glibly assume that Stonehenge was a temple. It might well have been, but we know nothing about the religions of those days and have no evidence either way. Nor is there any evidence for astronomer–priests. The builders of Stonehenge presumably had priests (nearly all peoples have religions, and most religions have officials of some sort), but there is no reason to suppose that the astronomers and the priests were the same persons.

There can be no doubt that the main axis of Stonehenge points deliberately to midsummer sunrise. The station-stones form a rectangle with sides parallel and perpendicular to the main axis. This would be a natural way to fit a rectangle into an edifice with an axis. But why fit a rectangle at all into an edifice that is essentially circular? The direction perpendicular to the main axis is, in fact, the direction of the southernmost moonrise. Did people notice this, and build a rectangle into Stonehenge to mark it? If so, why did they construct the station-stone rectangle rather than a cross-axis with a foresight like the heel-stone? Or did they build a rectangle for some other reason?

It has been suggested that the designers of Stonehenge deliberately looked for a site at which these two directions are at right angles, but we can safely reject this suggestion. If they were a widely traveled people, they might perhaps have looked for a site at which the northernmost and southernmost sunrises are perpendicular, but they would scarcely want a moonrise to be perpendicular to a sunrise. Moreover, if they had deliberately sought a site at which these two directions are perpendicular, they would have built a rectangular edifice, not a circular one with a rectangle unostentatiously incorporated.

If we look for further evidence for moonrise alignments, we notice the stone-holes D and F (see Figure 2.4) at just the right distance from the

main axis to give moonrise directions M_1 and M_2, 10° each side of the
axis. This could be a coincidence, because D and F are the width of the
avenue apart, and this could happen to give the required angle when seen
from the center. But between the main axis and the line through D is the
grid of small post-holes, in just the right position to keep track of the
small changes in the moon's rising-direction as it nears its northernmost
azimuth, as explained on pages 13 to 15 [35].

A rectangle fitted into a circle can have any proportions. How were the
proportions of the station rectangle chosen? One way would be to make a
diagonal point in some interesting direction; and, in fact, one diagonal of
the station rectangle points in the direction M_3. Once one diagonal has
been chosen, the other one is fixed, so it is no surprise that the other
diagonal points nowhere in particular.

So far we have accounted for all directions except midwinter sunrise.
This turns out to be given by stone-holes G (from 94) and H (from 93).
Why were the stones here set up to be seen from station-stones and not
from the center? And why were they placed on the circle through the
station-stones, and not a fair distance away like the heel-stone? One
possibility is that the designers of Stonehenge noticed not only that the
northernmost sunrise and southernmost moonrise are perpendicular,
but also that the northernmost moonrise and southernmost sunrise are
perpendicular. It is a geometrical fact that if G is on the circle the
directions 94-G and 92-G are perpendicular; consequently, if G is placed
to mark the southernmost sunrise from 94 it will automatically also
mark the northernmost moonrise (M_1) from 92. Similarly, H will mark
two directions (from 93 and 91). The archaeologists who worked at
Stonehenge were by no means sure that F, G, and H are stone-holes; the
holes are irregular in shape and might have been made by the roots of
trees or bushes now rotted away. It would, however, be a remarkable
coincidence if three bushes grew in just the right positions to give astronomical alignments [36].

To get some idea of whether the alignments could have come about by
chance we might turn to the science of statistics. What is the mathematical probability that a collection of point-to-point alignments could, by
sheer chance, come so close to so many astronomically interesting directions? First, we must decide how accurate we can reasonably expect an
alignment to be. Anyone who has moved a 100-kilogram stone (perhaps
building an oriental garden) will know how difficult it is to place it exactly
where it is wanted; and the stones of Stonehenge are much larger than
that. Gerald Hawkins suggested 2° accuracy [37]. R.J.C. Atkinson criticized this as being "arbitrary," but did not say what degree of accuracy
would not be arbitrary [38]. Hoyle suggested 0.3°, but this was for the
accuracy that could be obtained using poles, not megaliths [39]. There are
six interesting rising-positions, and the points on the horizon within 2° of
them cover altogether 24°, out of 180°. Therefore if we use Hawkins's

degree of accuracy, the probability that a random alignment scores a hit is 24/180. The next question is: How many alignments do we have? There are various ways in which we can count them. One way is simply to note that there are ten points (center, four stations, heel, D, F, G, H); joined up in pairs they give 45 directions. Between them they score 14 hits. Hawkins lumped his inner-Stonehenge results in with his outer-Stonehenge results, and claimed 24 hits out of 50 shots on a target covering 72° out of 360°. The probability of this happening by chance is about six in a million. Hawkins did not say how he counted the 50 lines. Atkinson maintained that he should have counted 111 lines, in which case the probability is about one in two. Each statistician got the result he wanted. If anyone else tries to apply the theory of statistics, the result will depend on how the shots are counted and how the target is defined. Both of these are highly arbitrary. Statistics will not settle the question.

Hawkins also suggested that the Aubrey holes could be used to keep track of the various cycles. If three black and three white stones were moved one hole at a time round the circle, he claimed, "the Aubrey holes could have been used to predict many celestial events," "an astonishing power of prediction could be achieved." and "the Aubrey hole computer could have predicted—quite accurately—every important moon event for hundreds of years." The "moon events" that he was referring to are eclipses. But, in fact, the most that the stones could have predicted is that when no stone is in a certain hole there will be no eclipse at the solstices or equinoxes of that year. Hoyle also had a theory about the Aubrey circle: according to him it is a giant protractor and was used to measure the directions of the sun, moon, and nodes [39]. However, such a protractor would have to be in the plane of the ecliptic (the Aubrey circle is not); moreover, the nodes are a mathematical fiction—they cannot be observed nor their directions measured. Both theories conflict with the archaeological evidence that the Aubrey holes were filled up soon after they were dug.

No advanced mathematical or astronomical theories were needed to set up the alignments at Stonehenge, only very painstaking and reasonably accurate observations. The marvel of Stonehenge lies not in its astronomy but in its construction; to erect such an edifice out of such massive stones without machinery of any kind is an impressive feat indeed.

Other Megalithic Structures

Other megalithic structures besides Stonehenge may mark interesting directions [40]. One is at Callanish, on the island of Lewis, at the rather high latitude of 58.2° [41]. There, thirteen upright stones, 3 or 4 meters high, form a somewhat flattened circle about 15 meters in diameter. In the middle is a larger stone, some $5\frac{1}{2}$ meters tall. A row of six stones runs south from the circle; a line of four stones runs west; a line of four stones runs somewhat north of east; and two parallel rows, ten stones in one and nine in the other, form an avenue 8 meters wide running slightly east of north.

Admiral Boyle Somerville, in 1912, noticed that a flat-sided stone to the north-east of the circle seemed to point to a tall stone to the south-west, and the line joining them gave a declination of $-28°10\frac{1}{2}'$, close to the southernmost declination of the moon. This was the first suggestion of an alignment to the *moon* in a megalithic structure. The inward direction along the avenue—from the far end toward the circle—corresponds to a declination of $-30.2°$, also close to the southernmost declination of the moon.

A more typical megalithic site is the one at Ballochroy, in Scotland, at latitude 55.7° [42]. Here three stones, about 4 meters, 4 meters, and 2 meters tall, respectively, 3 or 4 meters apart, stand in a straight line. This line points to a prominent peak in the direction corresponding to a declination of $-23.9°$, the declination of the midwinter sun, which grazes the peak as it sets. The stones are slab-sided, the middle one being particularly thin, and they are set at right angles to the line joining them. If we look along the slab sides of the stones we see another clearly marked peak, in a direction corresponding to a declination of 23.9°,

FIGURE 2.7. View to the northeast at Minard, on the shore of Loch Fyne, Scotland. The (re-erected) stone pillar and a notch on the horizon, as seen from a central boulder, mark the direction of midsummer sunrise. (Photograph E.C. Krupp, Griffith Observatory.)

that of the midsummer sun. (Because these two directions are at right angles to each other, the megalith builders could have erected a rectangular structure, perhaps like the station stones at Stonehenge; but they did not.) Figure 2.7 shows another site.

Alexander Thom, who has done nearly all the work on these megaliths, has investigated a couple of hundred alignments. Some point to the six rising-points of the sun and moon that we noted at Stonehenge, others to the corresponding setting-points, others to the rising- and setting-points of bright stars; and yet others to the rising- and setting-points of the sun at the equinoxes and the quarter-days (which are midway between the equinoxes and the solstices). In fact, there are some forty rising- and setting-directions that Thom considers to be of interest. But without written records there is no way of telling which, if any, of these alignments are deliberate and which, if any, are coincidences. For example, a line of nine stones (the Nine Maidens, in Cornwall) point to the spot on the horizon where the star Deneb rose, but the stones might have been set up in a straight line for some quite different reason and only by sheer chance point to one of the forty interesting points on the horizon.

Thom maintains that megalithic astronomers had solved the problem of being unable to observe the rising-directions of the moon corresponding to the maximum and minimum declinations (see page 16); they used certain stone grids, examples of which have been found in Scotland, to compute the theoretical northernmost rising-direction from the three most northerly directions actually observed. The mathematical technique needed for this is quite advanced: mathematicians describe it as approximating to a curve by the closest-fitting parabola.

Thom has also suggested a way of finding the exact date of the summer solstice in spite of the fact that the sun's declination is changing very slowly then [43]. We need to find a mountainside slightly steeper than the path of the setting sun, as in Figure 2.8(a). If the sun sets behind the peak a small part of it will appear for a moment in the notch a few minutes later, and an observer in the right position will see the gleam. The further right the sun is, the further left will the gleam be visible (Figure 2.8(b)). One procedure would be to arrange a row of observers as shown in Figure 2.8(b) a few days before the solstice. At sunset, those who saw the gleam in the notch raise their hands, and the leftmost position at which the gleam was visible is marked. By repeating this procedure on the next few nights, we can find the night on which the sunset is furthest right. This will be the sunset nearest the solstice.

If the solstice takes place at an instant when the sun is setting, the change in declination between then and the next sunset (or, for that matter, between then and the previous sunset) is 14″. At latitude 50° that corresponds to a change of 27″ in the sunset direction. If the notch is 35 kilometers from the observers, this angle corresponds to a distance of 4 meters along the line of observers; if the notch is further, it corresponds

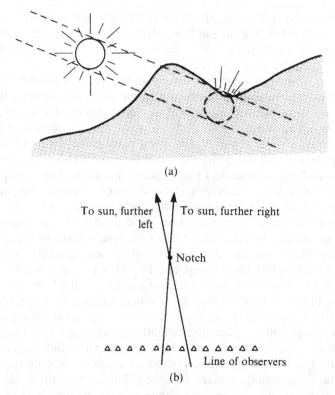

FIGURE 2.8.

to a greater distance (in fact, the distance along the line of observers is strictly proportional to the distance of the notch). If, however, the solstice is exactly midway between two sunsets, say between sunset on Wednesday and sunset on Thursday, then the sun will set in the same position on these two days. The change in declination between Thursday and Friday (or Tuesday and Wednesday) will be 29″, corresponding to a distance of 8 meters along the line of observers (for a notch 35 kilometers away). At latitude 60°, the same thing happens but the distances along the line of observers will be about 60% greater. All megalithic sites in Britain have latitudes between 50° and 60°.

The distances seem reasonable, but the method does depend on refraction not varying too much: if its change from one day to the next swamps the effect of the change in declination, the method will not work. As far as I know, no one has tried this method in practice, even though Thom has identified some suitable notches.

Thom's theories have inspired other workers to investigate megalithic alignments. Most of these later workers have cast doubt on Thom's alignments and rejected his statistics. However, Clive Ruggles has made

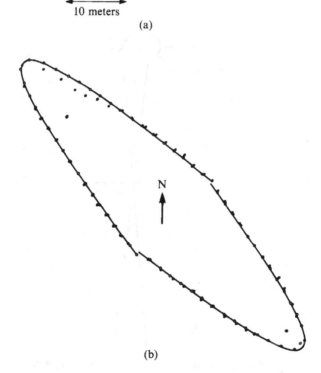

North

10 meters

(a)

N

(b)

FIGURE 2.9.

an extremely careful study of megaliths in the west of Scotland (the outer Hebrides, some of the inner Hebrides, and the nearby mainland coast) and has come to more favorable conclusions [44].

Ruggles found that the azimuths indicated by the megaliths are distributed in a very nonrandom way—they tend to point north or south and avoid pointing east or west. In fact, if we divide the complete circle into four equal quarters, 34 of Ruggles' best 36 alignments fall in the northern and southern quarters and only 2 in the eastern and western quarters. Ruggles considered various possible reasons for this—for example, the possibility that some alignments might be along a straight valley—but he found that geographical reasons cannot account for the nonrandomness. Another possibility—the one that interests us here—is that the azimuths

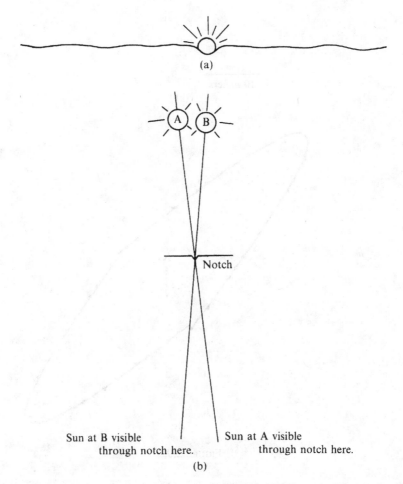

(a)

Notch

Sun at B visible   Sun at A visible
   through notch here.   through notch here.

(b)

FIGURE 2.10.

are nonrandom because the alignments were directed toward particular declinations. To show a north–south preference the alignments would have to avoid declinations close to zero and cluster just below ±35°. (At this latitude, 55°, with a flat horizon, the north point corresponds to a declination of 35°, and the declination decreases as we move away from north, the east and west directions corresponding to zero declination, and south corresponding to −35°. Declinations outside this range cannot be indicated by points on a flat horizon at this latitude.)

Ruggles found a significant tendency for the declinations to cluster round −31°, −29°, −25°, −22.5°, 28°, and 33°. The declination of the sun at midwinter was −24°, so the alignments indicating −25° (there are ten of them) might be toward midwinter sunset or sunrise, provided that the structures were not set up with great precision—1° is twice the diameter of the sun. The most southerly declination of the moon was −30°, so this

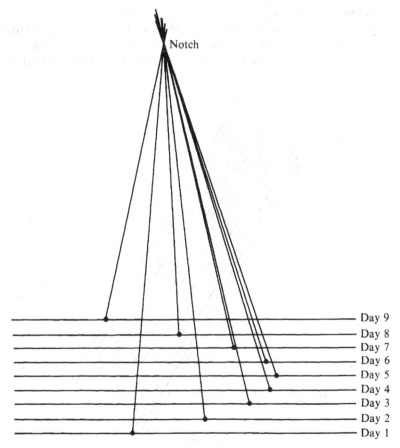

FIGURE 2.11.

might account for the clusters at −31° and −29°. The most northerly declination of the moon was 28°. The declination of 33° is too high to be given by the sun or the moon, but the alignments giving this might be ones along which the early astronomers looked the other way (i.e., toward the south, not toward the north); it is normally not possible to be sure which of the two directions along an alignment was the one actually intended for use. There seems to be no explanation of the cluster at −22.5°.

A rather unusual arrangement of stones is to be found at Ale, in southern Sweden [45]. It is shown in Figure 2.9(a). At first it was taken to be the outline of a Viking ship, and one of the stones, just inside the outline and near the "stern" was called the "steering-oar stone." Closer investigation shows that the stones lie pretty well on two parabolas: see Figure 2.9(b). The "ship" lies broadside on to the direction of mid-summer sunrise. These two facts make it plausible that the stones might have been used for marking the variation in the sun's rising-direction near midsummer day. The suggestion is as follows. The sun-watchers pick out some point on the horizon, perhaps a notch as in Figure 2.10(a), through which the sun can be seen as it rises. The sun will shine through the notch in slightly different directions on successive days (Figure 2.10(b)). The

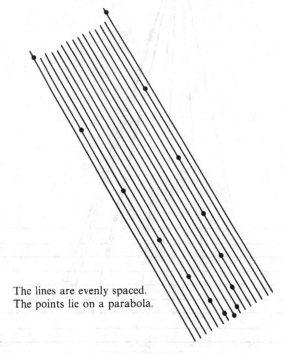

The lines are evenly spaced.
The points lie on a parabola.

FIGURE 2.12.

sun-watchers draw a number of equally spaced parallel lines at right angles to the direction of sunrise. A few days before midsummer they note the point on the first line from which the sun can be seen through the notch. The second day they mark the point on the second line, and so on, as shown in Figure 2.11. It is a mathematical fact that if the points were found in this way they would lie almost exactly on a parabola. The proportions of the parabola depend on how far away the notch is, and how far apart the parallel lines are. If this suggestion is correct, lines drawn through the stones parallel to the axis of the "ship" should be equally spaced, and the stones should be far from equally spaced along the parabola, (see Figure 2.12). Unfortunately, the monument was re-erected in 1916 and in any case the stones near the stern (or bow) of the ship, which are the ones that show the spacing most clearly, are irregularly placed and seem to have been disturbed. Presumably the other half of the "ship" was set up to make a pleasing symmetrical ship-like structure. It could not be used at midwinter sunset in the way described above because at midwinter the sun, as seen from the "ship," sets over the sea, and no notches are available.

CHAPTER 3

The Babylonians

Early Period

The earliest systematic stargazers were the Babylonians. Let us see what they actually recorded, according to tablets that have been found. There is a fascinating account of how the tablets were found and interpreted in Otto Neugebauer's *Exact Sciences in Antiquity* [46].

The earliest astronomical records were of times of moonrise and the date of the new moon, interspersed with notes on the weather. These go back to about 1800 B.C. A remarkably complete set of observations of risings and settings of Venus cover the 21 years (from 1702 to 1681 B.C.) [47] of the reign of Ammisaduqa, and even in these early days there were calculations as well as observations. Calculations are sometimes hard to distinguish from observations, but where there is a fairly complete series, calculations show up by their greater regularity. For instance, observations of a visible heliacal rising will be affected by mist or clouds. If the observer misses it one day, he will record it a day late. Calculations do not suffer from this kind of irregularity.

A slightly later series of tablets cover a period of about 1,000 years altogether, starting about 1100 B.C. Each tablet has three lists of stars: twelve "stars of *Anu*," twelve "stars of *Ea*" and twelve "stars of *Enlil*." The stars of *Anu* are close to the equator, those of *Ea* are somewhat to the north, and those of *Enlil* to the south. In each list the stars were more-or-less evenly spaced in the sky and serve as a calendar in much the same way as the Egyptian lists (see page 83). The early versions of the tablets were circular and divided into sectors (rather like a dartboard) and some had figures giving the length of daylight at various dates.

The Mul-Apin Tablets

One series of tablets are called *mul-apin*, after their opening words, see Figure 3.1. The earliest copy dates from 687 B.C. The tablets contain:

FIGURE 3.1. Copy of the mul-apin cuneiform tablet. (Photograph E.C. Krupp, Griffith Observatory.)

An extensive list of stars of *Enlil*, *Ea*, and *Anu*, and the dates of their heliacal risings.

A list of pairs of stars in opposite directions from the earth, so that one rises as the other sets.

A list of culminations. (A star is said to *culminate* when it reaches its highest point in its circle round the pole.) The tablets indicate the time of culmination by listing a star that is rising at the same time.

A list of stars and constellations in the moon's path.

The statement that the sun is in *Anu* in months XII–II and winds blow;
it is in *Enlil* in months III–V, the time of heat and harvest; it is in *Anu*
again in months VI–VIII and winds blow; it is in *Ea* in months IX–XI
and the weather is cold.

Rules for telling when a year needs a thirteenth month. For example, if
the moon is near the Pleiades on the first day of the year there will not
be a thirteenth month.

Tables showing when the shadow of a standard gnomon (a vertical rod
standing on a horizontal plate) has certain lengths.

The length of the night on various dates.

The statement about the sun being in *Ea*, *Anu*, or *Enlil* shows that the
Babylonians at this time saw how the height of the sun in the sky affects
the seasons. The fact that it is a separate item from the list of stars on the
path of the moon shows that the Babylonians did not yet realize that the
sun moves along the ecliptic.

From about the same date there are many astrological reports and one
of them, about 612 B.C., successfully foretold an eclipse of the moon [48].

The Zodiac

The list of stars and constellations on the path of the moon is a rough-
and-ready way of describing the ecliptic. Later the Babylonian astronomers reduced the list to twelve constellations roughly equally spaced
around the ecliptic. Later yet they divided the ecliptic into twelve precisely equal parts, which we call "signs of the zodiac," each named after
the constellation nearest it. (The signs of the zodiac are therefore not the
same as the constellations of the zodiac: a constellation is a group of
stars, whereas a sign is a twelfth part of a circle.) Thus a Babylonian
might describe a position on the ecliptic as "6° in *mul*," that is 6° further
east than the first point of *mul*. A degree, for which the Babylonian word
is *ush*, is one-thirtieth of a sign, and therefore 1/360th of a circle. The fact
that Babylonians used degrees does not mean that by some far-fetched
coincidence they used our units; *we* are using theirs (even after the
introduction of the metric system). The first list of stars which used the
signs of the zodiac to describe their positions is dated about 410 B.C.

Later, the Greeks used the signs of the zodiac. As far as we know, the
first Greek astronomers to use it were Meton and Euctemon, about 430
B.C. In fact, our word zodiac comes from the Greek word *zodion*. *Zodion*
could mean either a sign of the zodiac or a constellation of the zodiac;
when the Greeks wanted to be perfectly clear they used the word *dodeca-
temorion* [twelfth part] for a sign of the zodiac. The Latin translations of
the Greek names of the twelve chosen constellations are the ones familiar
to us. On the list on page 67 the signs are in the order in which the sun
passes through them (so that each is east of the one before) starting with

Latin	Babylonian		Greek	
Cancer	kushú, nangar, allul	[?]	Καρκινος (Karkinos)	[Crab]
Leo	a, ura	[lion]	Λεον (Leon)	[Lion]
Virgo	absin	[furrow]	Παρθενος (Parthenos)	[Virgin]
Libra	rin, zibanitu	[balance]	Χηλαι (Khelai)*	[Claws]
Scorpio	gír, gír-tab	[scorpion]	Σκορπιος (Skorpios)	[Scorpion]
Sagitarius	pa, pabilsag	[name of a god]	Τοξοτης (Toxotes)	[Archer]
Capricornus	másh, sahurmásh	[goat-fish]	Αιγοκενας (Aigokeros)	[goat-horned]
Aquarius	gu, gula	[?]	Ὑδροχοος (Hydrokhoos)	[water-pourer]
Pisces	zib, zib-me	[tails]	Ἰχθυες (Ikhthyes)	[fishes]
Aries	hun, luhunga, lu	[hired hand]	Κριος (Krios)	[ram]
Taurus	múl	[star]	Ταυρος (Tauros)	[bull]
Gemini	mash, mash-mash, mashtabba	[twins]	Διδυμοι (Didymoi)	[twins]

* Later replaced by Zygos [balance].

the sign that in classical times (about 500 B.C.) contained the sun at
midsummer.

Where precisely on the ecliptic were the signs placed? We can tell from
Babylonian tables for the length of daylight, which give the position of
the sun on the longest day. The Babylonians, it turns out, had two
systems. In one, the summer solstice was at 8° in *kushu* (and therefore the
winter solstice was at 8° in *mash*). In the other system the solstices were
at 10° in their signs. The Babylonians also measured the distance of the
moon and the planets from the ecliptic, using a unit called *she* which
literally means "barley corn" and is equal to 1/72 of a degree.

This is the system that the Greeks extended to cover the whole sky (see
page 32), measuring longitudes in signs and degrees and latitudes in
degrees north or south. Where exactly did the Greeks place the signs on
the ecliptic? Some, including most astrologers, placed them as in the first
Babylonian system, and traces of this placement can be found as late as
A.D. 1396 [49]. Most Greek mathematical astronomers had the solstices at
the beginnings of signs, and this system was used throughout the middle
ages. It is more convenient to measure longitudes in degrees only, from 0°
to 360°, starting with 0° in Aries as the zero point. Then

$$3° \text{ Aries } = 3°,$$
$$0° \text{ Taurus } = 30°,$$
$$7° \text{ Gemini } = 67°,$$
$$\text{etc.}$$

Observations

From about 700 B.C. onward we find clay tablets recording remarkably
sophisticated observations of the moon and planets, down to such details
as the direction in which the earth's shadow sweeps across the moon in an
eclipse. Over a thousand pieces of these tablets (they are very fragile)
have been unearthed. Fitting fragments together is like completing a
difficult but fascinating jigsaw puzzle, and has sometimes meant gathering
together parts of a tablet from different museums [50].

A typical tablet covers either the first or the second half of a year and
for each of the six months (or seven if an extra month is intercalated: see
page 20) the tablet records the following items:

the number of days in the preceding month;
the time between moonrise and sunset on the last day of the month on
    which the moon rises before sunset;
the times between sunset and moonrise on the next day;
the time between moonset and sunrise on the last day on which the moon
    sets before the sun rises;
the time between sunrise and moonset on the next day; and

the time between moonrise and sunrise on the last day on which the moon is visible.

(If any of the above is missed because of clouds, it is marked as "not observed" and an estimate is made.)

the longitudes of each of the planets (i.e., the signs of the zodiac in which they are situated);
the river level in Babylon; and
the price of barley, dates, sesame, etc.

Thus the tablets contain more than just astronomical data, they are really almanacs. They also include, where appropriate, the following items:

details of eclipses;
for each outer planet, the date of its first or last visibility, the date on which it starts or finishes retrograde motion, and the date on which it rises as the sun sets;
the date of first or last visibility of Venus or Mercury;
conjunctions of the moon or planets with stars near the zodiac;
the weather, if bad; and
interesting news.

The Babylonians had noticed the periodicity of the motions of the moon and the planets. For example, 71 years equals 65.01 synodic periods of Jupiter and 5.99 sidereal periods. Because these are very nearly whole numbers, the dates and longitudes of the interesting features of Jupiter's motion (as listed above) in any one year will be repeated almost exactly 71 years later. Therefore the data for Jupiter this year can be foretold by looking at the data for 71 years ago. A compilation for a planet used for this purpose is called a "goal-year text" (a somewhat awkward literal translation of the German Zieljahrtexte). Another interval that works well for Jupiter is 83 years; 59 years serves for Saturn, 79 and 47 years for Mars, 8 years for Venus, 46 years for Mercury, and 18 years for the moon. There are, in particular, data about eclipses of the moon arranged in 18-year groups, ranging from 731 B.C. to 317 B.C., and similar data for eclipses of the sun, ranging from 348 B.C. to 286 B.C.

The observational records were still being produced in the late period, right up to 50 B.C.

Sexagesimal Numerals

In order to follow the details in later Babylonian astronomy we need to understand Babylonian numerals. The system for writing numbers up to sixty was quite simple: a stroke represented one unit and a sideways vee-shape represented ten units. For example, 23 was written as

(i)
$$\langle\langle\,\vert\vert\vert$$

The Babylonians wrote numbers bigger than sixty in a system analogous to our decimal system but using *sixty* in the way that we use *ten*. To us

31

means three *tens* plus one. To a Babylonian

(ii)
$$\vert\vert\vert\ \ \vert$$

meant three *sixties* plus one, that is, 181. To avoid printing the strokes and vees I will transcribe them by modern numerals. The Babylonian numeral labeled (ii) above will be transcribed as

3,1

while (i) will be transcribed simply as 23.

Then, for example,

sexagesimal 1,10   = decimal 70,
sexagesimal 2,1,13 = decimal 7273 (2×60×60 + 60 + 13).

Addition in sexagesimals is like addition of degrees, minutes, and seconds. Compare the following two sums:

Modern		Babylonian	
	3° 41′ 23″		3,41,23
+	2° 18′ 41″		2,18,41
=	6° 0′ 4″		6, , 4.

Where the numerals were carefully written in columns, a blank space would do for zero. Elsewhere the Babylonians used a separation symbol.

This example shows how the Babylonians dealt with what we would treat as fractions. If a Babylonian were working in degrees and wanted an angle halfway between 3° and 4° (which we would denote by $3\frac{1}{2}$°) he would switch to a finer unit, namely a sixtieth of a degree. Now, 3 is 3,0 sixtieths and 4 is 4,0 sixtieths, and halfway between them is 3,30 sixtieths—just like our modern 3°30′. If this were not fine enough the Babylonian would switch to a sixtieth of a sixtieth, and so on. Thus our 3°31′41″, for instance, would be written by a Babylonian as 3,31,41 and interpreted as 3,31,41 sixtieths of a sixtieth of a degree.

A Babylonian had to rely on the context to tell which units were being used. This did not in practice cause any confusion. After all, in our

modern world, the Rover 2000, the Mercedes 200, and the Citroën 20 all
have engines of the same size and no one ever supposes that the Rover is
ten times the size of the Mercedes. However, it is sometimes convenient
to introduce a symbol analogous to the decimal point when we transcribe
Babylonian numerals for modern readers. If, for instance, we have identified 1,10,6 in a certain context as representing 1,10,6 sixtieths of a
certain unit, we can alternatively regard it as 60 + 10 + 6/60 of that unit,
and write it as 1,10;6 units, writing the "sexagesimal point," which turns a
comma into a semicolon, after the whole number.

There are two things that the Babylonians divided into 360 parts: the
circle and the day. They called 1/360 of a circle an *ush*, which therefore
means the same as our degree (or degree of arc) when used to measure
angles. Then 0;1 *ush* is a minute and 0;0,1 *ush* is a second of arc. I
translate *ush* as "degree" in this context.

The Babylonians also called 1/360 of a day an *ush*. I leave the word
untranslated in this context. One *ush* equals four of our minutes.

Late Period

Between about 300 B.C. and A.D. 75 Babylonian mathematical astronomy
reached remarkable heights. About three hundred tablets dealing with
mathematical astronomy have been found in various excavations. They
have been described a number of times, notably by Otto Neugebauer in
A History of Ancient Mathematical Astronomy, so I give only a brief
description here. Most of the tablets give numerical details of the motion
of the moon or a planet. These are called *ephemerides*, and consist of
rows and columns of figures with a few Babylonian words such as names
of months or of signs of the zodiac. Other tablets give some details of
how the figures were calculated: these might be called *instruction tablets*.
By putting together these details with calculations from modern data of
the positions of the sun, moon, and planets at specific times in the past,
and by carrying out some mathematical detective work on the figures in
the ephemerides tablets, modern astronomers have managed to interpret
nearly all the material.

The Sun

No Babylonian tablets deal directly with the sun. We have to deduce the
Babylonian theory of the sun's motion from details on the tablets dealing
with the moon.

A typical ephemerides tablet for the moon will deal with either the full
moon or the new moon, or perhaps will have figures for the full moon on
the front and the new moon on the back. The figures are arranged in rows
and columns. The first column is always a column of dates; in fact, it will

consist of the names of a number of consecutive months. Here follows a transcription of two columns from a full-moon tablet for the year 195 of the Seleucid era (i.e., 116 B.C.) [51]. The 3,15 is 195 in sexagesimals, and the months listed are the twelve months of the year 195 and the first month of the next year.

3 15	nisannu	9 7 30	gir
	ayaru	7 15	pa
	simanu	5 22 30	mash
	duzu	3 30	gu
	abu	1 37 30	zib
	ululu	52	hun
	tashritu	52	mul
	arahsamna	52	mash
	kislimu	52	kushu
	tebetu	52	a
	sabatu	52	absin
	adaru	37 30	rin
3 16	nisannu	28 45	rin

Gir, pa, etc. are the names of the zodiac signs, so the second column is obviously a list of longitudes. It gives, in fact, the longitude of the moon at the time of full moon in each month. The longitude of the sun then is easily found: it differs from the moon's longitude by exactly six zodiac signs, because at full moon the sun and the moon are in opposition.

We can now deduce the Babylonian theory of the sun's motion. For much of the year the sun advances one zodiac sign per month—this leaps to the eye because of the repeated 52's. For most of the rest of the year the sun advances $\frac{15}{16}$ of a sign per month, for example, from 9;7,30° in one sign to 7;15° in the next. (30° + 7;15° − 9;7,30° = 28;7,30°, which is $\frac{15}{16}$ of 30°.) This explains all the figures in the table except two: between the full moons in abu and ululu the sun covers 29;14,30° and between the full moons in sabatu and adaru it covers 29;45,30°. The simplest explanation for this is that the sun changes velocity from 1 sign per month to $\frac{15}{16}$ of a sign per month somewhere between the two full moons. It is not hard to calculate where the velocity changes: it changes from $\frac{15}{16}$ to 1 sign per month at absin 13° and changes back at zib 27°. Calculations with other tablets imply changes at exactly the same points. This is the kind of mathematical detective work that historians use to help them interpret the tables.

About half the Babylonian tablets are based on this fundamental assumption; the other half are based on a different but equally precise theory of the sun's motion. We call the two theories system A and system B.

Because months are not all of equal length, the fundamental assumption of system A does not quite mean that the sun moves with constant velocity from *absin* 13° to *zib* 27°. However, because the moon moves round the ecliptic much faster than the sun, the variation in the moon's velocity has much less effect on the length of the month than has the variation in the sun's velocity. Therefore, the months in which the sun moves with its fast velocity are approximately equal (and the months in which it moves with its slow velocity are also approximately equal). The approximation is good enough for the calculations which follow. In any case, the Babylonians presumably did not believe that the sun suddenly changes velocity twice a year, but regarded the whole system as a reasonable approximation that worked in practice. The important point is that the Babylonians had discovered the vitally important fact that the velocities of the sun and the moon vary.

When the position of the sun on the ecliptic is known, the length of the day from sunrise to sunset can be found. The Babylonians found it as follows. First, they found the positions of the sun at the equinoxes: these were *hun* 10° at the spring equinox and *rin* 10° at the autumn equinox. An *ush* is 1/360 of a day; therefore at the equinoxes day and night are each 180 *ush*.

The time from sunrise to sunset is the same as the time from sunrise to the moment when the point of the ecliptic diametrically opposite to the sun rises. The different zodiac signs, although equal in length, take different amounts of time to rise, because of the change in angle between the ecliptic and the horizon. We can see this in Figure 3.2, in which *A* and *B* are two points one degree apart on the ecliptic. *CB* is drawn in the direction in which *B* moves as it rises, i.e., perpendicular to the direction toward the celestial pole. *CB*, measured in degrees, shows how far the celestial sphere has to turn for the arc *AB* to rise, and therefore (converted to time at the rate of 360° per day, i.e., one *ush* per degree) shows how long *AB* takes to rise. The length of *BC* clearly depends on the angle

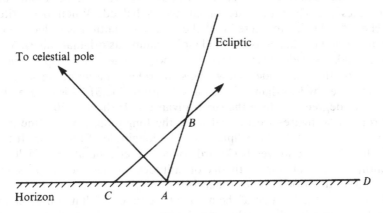

FIGURE 3.2.

between the ecliptic and the horizon, and is greatest when this angle is greatest, which will be at the spring equinox.

In the Babylonian theory the rising-times in *ush* are as follows: (What we call *hun* for convenience is really the arc of the ecliptic from *hun* 10° to *mul* 10° and so on) [52].

hun	*mul*	*mash*	*kushu*	*a*	*absin*	*rin*	*gir*	*pa*	*mash*	*gu*	*zib*
20	24	28	32	36	40	40	36	32	28	24	20

The length of the day corresponding to, say, *mul* 10° is the time taken for half the ecliptic, starting at *mul* 10° to rise, i.e., 24 + 28 + 32 + 36 + 40 + 40 = 200 *ush*. The following table gives the lengths of the day in *ush* when the sun is at 10° in the successive signs of the zodiac:

hun	*mul*	*mash*	*kushu*	*a*	*absin*	*rin*	*gir*	*pa*	*mash*	*gu*	*zib*	*hun*
180	200	212	216	212	200	180	160	148	144	148	160	180

The Babylonians assumed that the length of the day changes uniformly between these tabulated values.

The Moon

Latitude

The Babylonian theory of the moon's latitude has been deduced from the ephemerides tablets by mathematical detective work rather like the calculations we made for the sun's motion.

The latitude is measured in *she*, which means "barley-corn"; one *she* is one seventy-second of a degree. In a month in which the sun advances by 30° the moon's latitude changes at a rate of 252;31,24 *she* per month until it reaches 2°. Then the rate of change is halved. When the latitude reaches 6° it begins to decrease at the same rate until it is reduced to 2°, when the rate of change is doubled (so regaining its original value). When the latitude is reduced to zero the moon crosses the ecliptic and its latitude on the other side follows the same rules. In a month in which the sun advances by less than 30° the basic figure of 252;31,24 is reduced by 8 for every degree by which the sun's advance falls short of 30°.

From these figures we can calculate the longitude at which the moon crosses and recrosses the ecliptic, i.e., the longitude of the node. It turns out that the node moves backward round the ecliptic at 1;33,55,30° per month, so the Babylonian theory of latitudes implies the "regression of the nodes" described earlier (page 14).

If we know the latitude of the moon at the time of full moon we can tell whether or not it will be eclipsed. The Babylonians knew that an eclipse cannot occur when the moon is more than 17;24 fingers from the ecliptic (a finger is one-twelfth of a degree). If the full moon is x fingers nearer than this to the ecliptic, then x fingers of it will be eclipsed. The Babylonians recorded the magnitudes of eclipses, in fingers, on the ephemerides tablets.

The moon goes through a complete cycle of latitude (e.g., from greatest value to least and back to greatest) in one latitudinal period, which is a little less than a month. Thus in a month it goes through a complete cycle and a little bit more. It is the "little bit more" that is calculated by the rules that we have been describing, because the tablets record the latitudes only at monthly intervals.

The Lengths of the Months

The tablets are also concerned with calculating the times and dates of new moon (and full moon). This gives the length of the month, because each month starts with a new moon. Theoretically, the basic variable here is the velocity with which the moon advances round the ecliptic. However, this velocity could not be measured directly with anything like the accuracy needed. Nor can one measure the position of the new moon or full moon. When the moon is new it cannot be seen, and when it is full it remains full (as far as the eye can judge) for some time, and its position on the ecliptic is changing all the while. But there is one occasion when the time of full moon can be found quite accurately—when the moon is eclipsed. The instant halfway between the start and finish of the eclipse is the precise time of full moon. We cannot measure the length of any one month in this way because the moon is never eclipsed twice running (see page 19) but if we time two eclipses x months apart we can find the precise total of these x months.

We saw on page 19 that 223 months (i.e., one saros) is almost exactly a whole number of latitudinal half-periods of the moon. This suggests that one saros after an eclipse there is a good chance of another. Will the variation in the moon's velocity upset this? The time taken by the moon's velocity to go through a complete cycle of changes (e.g., from maximum to minimum and back again) is called the "anomalistic period" of the moon; the name comes from the practice of calling any departure from regularity an anomaly. In each anomalistic period the moon's velocity goes through the same cycle of changes, and so the moon advances the same distance round the ecliptic. It turns out that a saros is almost exactly a whole number of anomalistic periods: a saros averages 6,585.32 days, and 239 anomalistic periods add up to 6,585.45 days. Therefore the distance by which the moon advances round the ecliptic in a saros varies very little: the variation is the difference between the greatest and least distances the moon can cover in the fraction of a day by which the saros falls short of 239 anomalistic periods. Therefore the variation in the moon's velocity will scarcely affect the chance of an eclipse repeating after a saros.

To sum up: the length of the month is what the Babylonians wanted to calculate. Its variation is caused by variations in the moon's velocity. Neither of these could be measured directly. But the variation in velocity causes a change in the length of the saros, and this could be measured. The Babylonians therefore based their calculations on the length of the saros. The first column after the dates in system A tablets is a column of figures which have been recognized after much painstaking detective work as the length of the saros, and whose construction is precise and simple.

The Babylonian theory is as follows. From one new moon to the next (or one full moon to the next) the length of the saros increases by

$$2;45,55,33,20 \; ush$$

up to a maximum of

$$6,585 \; \text{days plus} \; 137;4,48,53,20 \; ush.$$

(It is only the extra *ush* that are tabulated: the 6,585 days are taken for granted.) When the saros length reaches its maximum it starts to decrease at the same rate until it reaches a minimum of

$$(6,585 \; \text{days plus}) \; 117;47,57,46,40 \; ush.$$

It then increases at the same rate to the same maximum and so on. The Babylonians presumably chose the maximum, minimum, and rate of change to agree with their measurements, though we have no details of how they did this.

This is an example of a variable that changes by a constant amount from one entry to the next. Variation at a constant rate back and forth between a fixed minimum and a fixed maximum is called *zig-zag* variation. The Babylonians based a great deal of their theory on it. We met another example of it earlier: the table of rising times on page 74.

If at the instant T the moon is $x°$ ahead of the sun on the ecliptic, the time the moon takes to pass the sun 223 times and end up $x°$ ahead is the saros length corresponding to T. On most tablets, it is the saros length *at monthly intervals* that appears, and it is only to these that the above rules apply. (There is one tablet that gives the saros length at daily intervals [53].)

The saros length for the new moon has the same minimum, maximum, and rate of change as the saros length for the full moon, but one has its minimum where the other has its maximum. We can see a reason for this if we ignore, for the moment, the slight change in the length of the month, and assume that all months are the same length. The saros length varies in zig-zag fashion, and its period is the anomalistic period of the moon. Figure 3.3(a) shows its graph. $A, B, C \ldots$ indicate successive in-

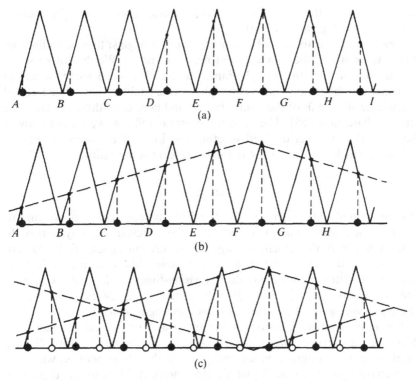

FIGURE 3.3.

stants when the saros length is a minimum. Black dots indicate the times
of successive full moons. Because we are assuming that all months are
equally long, these dots are equally spaced. Because a month is a little
longer than an anomalistic period, they are a little further apart than A,
B, C, etc. The points on the graph, vertically above the dots, show
the saros length for full moons. These are the lengths tabulated in a
Babylonian full-moon table, and so the dashed line in Figure 3.3(b) shows
the graph of saros length that would be deduced from a full-moon table.
It has the same maximum and minimum as the underlying complete saros
length function but a much longer period.

Each new moon comes midway between two full moons. Figure 3.3(c)
shows the full moon and new-moon zig-zags on the same diagram. This
shows clearly that one has a maximum where the other has a minimum.

The relation between the moon's velocity and the length of the saros in
the Babylonian theory was quite simple. One instruction tablet gives
a procedure for computing the velocity in degrees per day which is
equivalent to the following formula:

$$\text{velocity} = 15\tfrac{9}{16} - 27x/6{,}400,$$

where *x ush* is the entry in the saros tables. This makes the moon's velocity a zig-zag variable [54].

The final task is to find when the new month begins; that is, to find the first day on which the new crescent moon is visible. We do not know all the details of how the Babylonians did this. We have some details in instruction tablets, but only two ephemerides tablets have been found containing the result of the computation, and neither of them contains the intermediate steps [55]. The most important detail that we do not know is how the Babylonians decided whether the interval between sunset and moonset was long enough for the new crescent to be visible.

System B

As mentioned earlier (page 72), the Babylonians had two systems of mathematical astronomy. The one I have been describing is system A. In system B the sun's velocity changes gradually (in zig-zag fashion), not suddenly, and there is no saros length table. Many of the underlying figures are different: for example, the rising-times of the first six signs of the zodiac are 21,24,27,33,36,39 *ush*, instead of 20,24,28,32,36,40; and the equinoxes and solstices are at 8° in their signs instead of 10°.

For more details of system B the interested reader can consult Otto Neugebauer's *History of Ancient Mathematical Astronomy*, Book II. The mathematics in system B is less complicated than in system A, while the underlying figures are on the whole more accurate. It is likely, therefore, that system B was the later of the two to be invented, but all we can say for certain is that system A was used at least from 263 B.C. to 14 B.C. and system B at least from 251 B.C. to 68 B.C.

Lengths of Fundamental Periods for the Moon

In system B, we can read off the average length of the month from the table of month lengths. The result is

$$1 \text{ month} = 29;31,50,08,20 \text{ days},$$
$$(= 29.5305941 \ldots \text{ days in modern decimals}).$$

This figure is interesting because it was known to the Greek astronomer Hipparchus. The modern value is 29.530589 . . . days.

Two more system B results that Hipparchus knew are

$$269 \text{ anomalistic periods} = \quad 251 \text{ months},$$
$$5{,}923 \text{ latitudinal periods} = 5{,}458 \text{ months}.$$

The first is found from tables of the moon's velocity. The second is found from tables of eclipse magnitudes, which have the same period as the moon's latitude.

The Planets

Beside the moon, the Babylonians kept track of the planets. They calculated the dates of:

the instants when the planet starts and ends its retrogression;
the first visible *heliacal rising*;
the last visible *heliacal setting*; and
opposition.

They also computed the position of the planet on the ecliptic at these instants.

From the tablets we can deduce the underlying theory. Some use system A: the ecliptic is divided into a "fast arc" and a "slow arc." This is inevitable if the sun is assumed to move according to system A, because the phenomena listed above are tied in with the motion of the sun. The technical term for the angle between a planet and the sun as measured from the earth is *elongation*. The Babylonians knew that the start of retrogression always takes place at the same elongation; so does the end of retrogression; so do the heliacal rising and setting. And at opposition the elongation of a planet is necessarily 180°.

Some tablets divided the ecliptic into more than two arcs. One Jupiter tablet, for instance, divided the ecliptic into four arcs: one fast, one slow, two medium. The velocity on the fast arc was:

from heliacal rising: 15′ per day for 30 days,
      then   8′ per day for 3 months,
      then   5′ per day retrogression for 4 months,
      then   7′40″ (forwards) per day for 3 months,
      then   15′ per day for 60 days.

On the slow arc these velocities were multiplied by $\frac{5}{6}$; on each medium arc, by $\frac{15}{16}$. This takes us to the next heliacal rising, and the cycle starts again.

One Mars tablet divided the ecliptic into six arcs. Each arc consisted of two signs of the zodiac, and was subdivided into intervals as follows:

zib & hun	mul & mash	kushu & a	absin & rin	gir & pa	mash & gu
16 intervals	24 intervals	36 intervals	27 intervals	18 intervals	12 intervals

The distance round the ecliptic between two occurrences of the same phenomenon was 18 intervals. Suppose, for instance, that Mars began to retrogress at 26°40′ *pa*. This is the end of the seventeenth interval in the *gir*-and-*pa* arc. Eighteen intervals on from here takes us to the end of the fifth interval in arc *zib*-and-*hun*. This is 18°45′ *zib*; and this is where the next retrogression started [56].

For Mercury and Venus, the tablets record:

the first visible heliacal rising;
the start of retrogression;
the beginning of the period in which the planet is too near the sun to be
    seen;
the end of this period;
the end of retrogression; and
the heliacal setting.

The velocity of Venus does not vary much, and there is a tablet (from Uruk) in which the ecliptic is not divided into fast and slow arcs. For neither Mercury nor Venus are any tablets based on system B.

The period between two successive occurrences of the same phenomenon is the *synodic period* of the planet (see page 21). The time taken to go completely round the ecliptic is the *sidereal period*. From the tablets we can deduce the following average figures:

Jupiter	36 sidereal periods =	391 synodic periods =	427 years,
Saturn	9 sidereal periods =	256 synodic periods =	265 years,
Mars	151 sidereal periods =	133 synodic periods =	284 years,
Venus	1,152 sidereal periods =	720 synodic periods =	1,152 years,
Mercury	46 sidereal periods =	145 synodic periods =	46 years,
or	388 sidereal periods =	1,223 synodic periods =	388 years.

For example, in the Mars tablet quoted above the total number of intervals is 133. Thus after 133 synodic periods the start of retrogression has gone round the ecliptic 18 times and is back where it started. But in one synodic period, Mars goes once round the ecliptic plus the extra distance from one position for the start of retrogression to the next. Thus Mars has traveled 133 + 18 = 151 times round the ecliptic. Therefore 133 synodic periods = 151 sidereal periods. This is how the figures in the first two columns are found.

The principle that each phenomenon takes place at a fixed elongation means that in one synodic period the sun goes round the ecliptic the same distance as the planet plus a whole number of revolutions. For Jupiter, Saturn, and Mars this whole number is 1; for Venus and Mercury it is zero. Thus in 36 sidereal periods of Jupiter the total advance made by the sun is the total advance made by Jupiter plus one revolution for each synodic period, a total of 427 revolutions. This takes, of course, 427 years. That is how the figures in the last column are obtained.

This explains how we deduce these averages from the tablets. The Babylonians worked the other way round; they used these averages as basic data and constructed the tablets from them. How did they find the basic data? Probably by observing two occurrences of the same synodic phenomenon—say the start of the retrogression of Jupiter—that occur (a) a whole number of years apart, and (b) at nearly the same point of the ecliptic, so that the number of sidereal periods that elapse between the two observations is nearly a whole number. It is easy to count the number of periods that elapse. For example, one tablet for Jupiter shows (approximately)

6 sidereal periods = 65 synodic periods = 71 years,

and another shows (approximately)

1 sidereal period = 11 synodic periods = 12 years [57].

The error in the sidereal periods can be found from the position on the ecliptic. The first tablet showed that Jupiter traveled 6° more than a whole number of revolutions in the 71 years, while the second shows that Jupiter traveled 5° less than a complete number of revolutions in 12 years. Thus in 5×71 + 6×12 years (i.e., 427 years) Jupiter will travel almost exactly a whole number of revolutions. This combination of (i) and (ii) gives the basic data for Jupiter (5×65 + 6×11 = 391, 5×6 + 6×1 = 36).

CHAPTER 4

The Egyptians

The dynasties of ancient Egypt were roughly contemporary with the civilizations of Mesopotamia. But although Egypt was prominent in military might, architecture, art, literature, and some sciences such as medicine, it gave very little weight to astronomy. The Egyptians' lack of interest in astronomical matters is shown clearly by a "catalogue of the universe" compiled by Amenḥope about 1100 B.C. [58]. It lists only five constellations, of which two can be identified as Orion and the Great Bear, and does not even mention Sirius nor list the planets.

Astronomy fares a little better in a document of 300 B.C.—very late in Egyptian history (the first dynasty is dated about 3100 B.C., and ancient Egyptian history effectively ended with the conquest of Egypt by Alexander the Great in 332 B.C., though, of course, Egyptian culture did not cease abruptly at that date). The document I am referring to is a eulogy engraved on the base of a statue of a man named Harkhebi [59]. It describes him as observing "everything observable in heaven and earth" including the culmination of "every star in the sky," foretelling the heliacal rising of Sirius, and knowing the north–south movement of the sun. We do, in fact, have some evidence from alignments of temples that the Egyptians noted the rising-directions of the sun at the solstices (see note 8), and the heliacal rising of Sirius marked the start of the Egyptian new year.

In very early times the Egyptians based their months on the phases of the moon, starting each month when the old moon became invisible, which seems a little more awkward than starting the month when the new moon became visible, as most early civilizations did. As we have seen, there are twelve months plus about eleven days in a year. Thus most Egyptian calendar years consisted of twelve months, which were divided into three seasons—inundation, growth, and harvest—of four months each. If the heliacal rising of Sirius occurred in the last eleven days of the twelfth month the Egyptians inserted a thirteenth month. This kept the calendar year in step with the seasons. However, quite early on the Egyptians abandoned this calendar and introduced a 365-day calendar (described on page 20) dependent only on counting the days and not on observing the sun, moon, Sirius, or any other astronomical phenomena.

The Egyptians listed 36 groups of stars, usually called decans, each decan having its heliacal rising some ten days later than the previous one. Each decan is about 10° further round the celestial sphere than the previous one (i.e., has a right ascension greater by 10°), and on any one night it will rise some 40 minutes later. Consequently, the Egyptians could use the decans to tell the time at night. Tables to help them do this have been found on (of all places) the insides of coffin lids.

The first column of one of these tables shows the twelve decans that rise on one particular night, in the order in which they rise. The twelve decans will take about eight hours to rise and even allowing for the fact that darkness sets in some while after sunset and the dawn sky lightens some while before sunrise this is an unexpectedly short night. However, the Egyptians have left no explanation for this and modern commentators have not ventured any. The second column of the table shows the decans as they rise ten days later. The time of sunset does not vary greatly in ten days; consequently each decan will be rising some 40 minutes (relative to sunset) earlier than before, and the first decan in the first column will now be rising too early to be seen. The original second decan will now be the first to be seen rising, and so on. In fact, each decan will be positioned one place higher in the second column than it was in the first column. The same reasoning applies to all the columns, which between them cover a year at ten-day intervals. Therefore the occurrences of each decan occupy a diagonal line across the table, and this fact gives the tables their usual name: diagonal calendars.

Compared with Babylonian astronomy, these accomplishments amount to disappointingly little. No wonder that R.A. Parker, in his contribution [60] to the 1974 symposium *The Place of Astronomy in the Ancient World*, commented, "What are we to talk about when the subject is Ancient Egyptian astronomy? Little enough, it seems," though he did go on to say "but that little is not devoid of interest."

CHAPTER 5

The Chinese

Introduction

While the Babylonians were developing their astronomical systems,
another great civilization, the Chinese, was doing the same thing. The
earliest records of Chinese astronomy that have come down to us are
inscriptions on the famous oracle bones from Anyang (1500 B.C. onward).
Perhaps the most fascinating records on these bones are of "guest stars"—
novas, supernovas, and bright comets.

Chinese astronomy was a government service; we know that it was
nationwide because some eclipses that were not visible in the capital were
reported from outlying districts. Officialdom did not always understand
astronomy perfectly and there is a story in the *Shu Jing* (an early semi-
legendary history whose date is not known for certain) of an expedition
sent to punish astronomers for failing to prevent an eclipse. Even later,
in historical times, Chinese astronomers were charged with somewhat

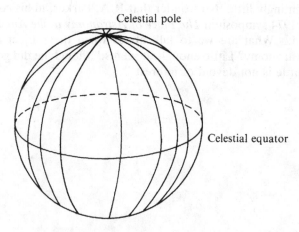

FIGURE 5.1. Division of the celestial sphere into *xiu*.

bizarre duties: for instance, "to deduce from the twelve winds the state of harmony of heaven and earth."

To describe the position of a star in the sky, the Chinese divided the celestial sphere into 28 parts, like the segments of an orange, stretching from pole to pole (see Figure 5.1). These segments were called *xiu*; they were not equally wide, and each was named after a constellation that it contained. The names are as follows:

Jiao	[horn]	*Nu*	[girl]	*Bi†*	[net]
Gang	[neck]	*Xu*	[emptiness]	*Zui*	[tortoise]
Di	[root]	*Wei**	[roof]	*Shen*	[investigator]
Fang	[room]	*Shi*	[house]	*Jing*	[well]
Xin	[heart]	*Bi†*	[wall]	*Gui*	[ghost]
Wei*	[tail]	*Kui*	[leg]	*Liu*	[willow]
Ji	[basket]	*Lou*	[bond]	*Qi xing*	[seven stars]
Nan tou	[southern ladle]	*Wei**	[belly]	*Zhang*	[net]
Niu	[ox]	*Mao*	[station]	*Yi*	[wing]
				Zhen	[chariot platform]

* These are different words in Chinese, written with different characters.
† So are these.

To specify the position of a star the Chinese cited the name of the *xiu* in which it was located, together with its angular distance from the pole. They measured angles in *du*'s: there are $365\frac{1}{4}$ *du*'s to a complete circle, so a *du* is a trifle less than a degree. For a more exact description they could say how far the star is from the edge of the *xiu*. For instance, the southernmost star of the constellation *Dongxian* is 103 *du* from the pole and 2 *du* forward from the beginning of *xin*. Three lists of stars made between 370 B.C. and 270 B.C. no longer exist, but before they were lost they were used (about A.D. 300) to make a map of the heavens. They contained between them about 1,464 stars.

By 20 B.C. the Chinese knew how eclipses were caused, but some astronomers resisted the explanation. Wang Chong, in the first century A.D., asked how the moon could possibly cause an eclipse of the sun in view of the fact that the moon is Yin and therefore weak and the sun is Yang, and therefore strong. He preferred to believe that it was "in the sun's nature" to grow dim at the time of an eclipse. By 8 B.C. the Chinese could predict eclipses by using the 135-month period; and by A.D. 206 they could predict eclipses by analyzing the motion of the moon. By A.D. 390 they could predict how much of the moon would be in shadow.

The Chinese observed sun-spots as early as 29 B.C., looking through slices of jade. (In the west, only two sightings of sunspots are known—both from fourteenth-century Russia—before the seventeenth century [61].)

The earliest known detailed description in China of the motion of the planets dates from about 90 B.C. The figures below are from the *Si feng* almanac of about A.D. 100 [62].

Saturn	Days	86	33		102	33	86	38 + 2,163/36,384
	du	6	0		−6	0	6	6 +

Jupiter	Days	58	58	25	84	25	58	58	32 + 14,641/17,308
	du	11	9	0	−12	0	9	11	4 +

Mars	Days	184	92	11	62	11	92	184	143 + 1,872/3,516
	du	112	48	0	−17	0	48	112	110 +

The tables start when the planet first becomes visible. Jupiter, for instance, moves forward at a certain rate for 58 days, covering 11 *du*, then a little slower for another 58 days, covering only 9 *du*, then remains stationary for 25 days, then retrogresses 12 *du* in 84 days, and so on. The last column covers the period when it is invisible and the odd fraction is there to make the total synodic period correct: this almanac took 4,327 synodic periods of Jupiter as equaling 4,725 years, giving an average synodic period of 398 + 14,641/17,308 days.

Venus and Mercury each have two periods of invisibility: an asterisk marks the second one:

Venus	Days	91	91	46	8	10	10*	10	8	46	91	91	82 + 562/23,320
	du	113	106	33	0	−6	−8	−6	0	33	106	113	100 +

Mercury	Days	20	9	2	1	18*	1	2	9	20	33 + 41,978/47,632
	du	25	8	0	−1	−14	−1	0	8	25	65 +

In A.D. 604 Liu Chuo developed an arithmetical technique, very similar to the Babylonian system B in principle, for dealing with the irregularity of the sun's motion.

In the ninth century A.D. the Chinese compiled elaborate tables of planetary positions along the lines of Babylonian tables; the idea might very well have come to China from Babylonia via Persia.

Chinese Units

In order to follow Chinese accounts in any detail it is necessary to know something about Chinese units of length and time, and Chinese methods of dating.

Units of Length

For small lengths the Chinese had a decimal system as follows:

1 *zhang* = 10 *chi* = 100 *cun* = 1,000 *fen*,   etc.

The important units are the *chi*, sometimes called the Chinese foot, and the *cun*, sometimes called the Chinese inch. In translations I will write

1 *zhang*   2 *chi*   3 *cun*   4 *fen*   as   12.34 *chi*,

and so on. The lengths of these units varied slightly in the course of history. For instance, the *chi* was 25.46 cm in the Tang dynasty (around A.D. 700) and 24.37 cm in the Yuan dynasty (around A.D. 1300). Do not be alarmed if you find in an encyclopedia the substantially bigger value 35.8 cm; this is a late-nineteenth century *chi* standardized for the purposes of international trade.

For longer distances the unit is the *li*, which was normally 1,800 *chi*. This would make 1 *li* equal to 0.44 km in Tang times.

Dates

The Chinese month started with the new moon. The astronomical year started with the month that contained the winter solstice, but from 104 B.C. onward the civil new year started two months later. (If there is a Chinatown in your city you will probably have noticed that the Chinese celebrate their new year in February.) Each Chinese year overlaps two of our years. For example, the Chinese year that started in February of A.D. 1000 covered most of A.D. 1000 and the first half-dozen weeks of A.D. 1001. For convenience I call this the year corresponding to A.D. 1000.

Chinese years were numbered in each reign-period, a reign-period being given an arbitrary name by the emperor. For example, the *Kaiyuan* reign-period started in A.D. 712. Then A.D. 713 is the second year of *Kaiyuan*, A.D. 724 is the thirteenth year of *Kaiyuan*, and so on. This reign-period lasted 29 years, and so its last year was A.D. 740. Then A.D. 741 was the first year of the next reign-period, *Tianbao*. There is no difficulty in identifying Chinese dates: tables are available giving the Western equivalent of every Chinese year and vice versa [63].

Another way of naming Chinese years is by using a sixty-name cycle. This cycle is organized as follows. There are two sequences of Chinese characters (see Figure 5.2). One sequence consists of twelve *zhi* [branches]:

1	2	3	4	5	6	7	8	9	10	11	12
zi	chou	yin	mao	chen	si	wu	wei	shen	yu	xu	hai

子	zi	甲	jia
丑	chou	乙	yi
寅	yin	丙	bing
卯	mao	丁	ding
辰	chen	戊	wu
巳	si	己	ji
午	wu	庚	geng
未	wei	辛	xin
申	shen	壬	ren
酉	yu	癸	gui
戌	xu		
亥	hai		

The
twelve
zhi

The
ten
gan

FIGURE 5.2.

The other sequence consists of ten *gan* [trunks]:

I	II	III	IV	V	VI	VII	VIII	IX	X
jia	*yi*	*bing*	*ding*	*wu*	*ji*	*geng*	*xin*	*ren*	*gui*

Each of the sixty names consists of a *gan* followed by a *zhi*. The first
name is *jia-zi* (I,1), the second is *yi-chou* (II,2), and so on. When either
sequence comes to an end it starts again at the beginning. Thus after
(X,10) comes (I,11), then (II,12), then (III,1), and so on.

Today, the Chinese seem to use only the *zhi*, not the *gan*. Each year, at
the time of the Chinese new year, they celebrate the year of the rat (*zi*)
or year of the bull (*chou*) or whatever the new *zhi* is.

Figure 5.3 is a table from which the number in the cycle of any *gan-zhi*
combination can be found, and vice versa.

Most Chinese records give the *gan-zhi* name of the year, as well as its
reign-period number. This is a great help. It is not, for example, possible
to tell the number of years between the thirteenth year of *Kaiyuan* and
the third year of the next reign-period without knowing how long the
Kaiyuan reign-period lasted; but if *Kaiyuan* 13 is given as *wu-chen* (5 in
the cycle) and *Tianbao* 2 as *bing-xu* (23 in the cycle), then the interval
between them is clearly 18 years.

Chinese months are numbered, not named, starting afresh each year.
Twelve months contain about 354 days, less than a year; thirteen months
are longer than a year. As a result some years contain twelve months,

	jia	yi	bing	ding	wu	ji	geng	xin	ren	gui
zi	1		13		25		37		49	
chou		2		14		26		38		50
yin	51		3		15		27		39	
mao		52		4		16		28		40
chen	41		53		5		17		29	
si		42		54		6		18		30
wu	31		43		55		7		19	
wei		32		44		56		8		20
shen	21		33		45		57		9	
yu		22		34		46		58		10
xu	11		23		35		47		59	
hai		12		24		36		48		60

FIGURE 5.3.

some thirteen. This, together with the fact that some months contain 29 days, some 30, means that the Chinese calendar is quite irregular. It is, however, completely known, and the tables mentioned above will convert Chinese dates unambiguously into European dates and vice versa [63]. For example, if we look up *Kai-yuan* 13 x 24 (i.e., the thirteenth year of the *Kai-yuan* reign-period, tenth month, twenty-fourth day) we find A.D. 724, December 3. The extra months in a thirteen-month year could be inserted anywhere. If in a certain year it was inserted after month 3, the months of that year would be named

1   2   3   *jian* 3   4   5   6   7   8   9   10   11   12

jian literally means "among"; in this context it is best translated as "intercalated."

The cycle of sixty names used for years is also used for days, where it is equally helpful. It is not possible to tell the number of days between ix 15 and xi 3 of the same year without knowing how many days there were in months ix and x, but if ix 15 is given as the seventh in the cycle and xi 3 as the fifty-fourth then xi 3 is 47 days after ix 15.

Time of Day

Chinese astronomers divided the day into 100 *ke*. Thus a *ke* was a trifle less than a quarter of an hour. For everyday purposes, however, the day was divided into 12 *shi*. Thus a *shi* was 2 hours. These double hours were not numbered but given the names of the twelve *zhi*.

If a time calculated by astronomers in *ke* is to appear in a civil calendar, it is given in a mixture of double hours and *ke*. For example, the winter solstice of A.D. 1277 occurred 32 *ke* after midnight on a day which is the fortieth (*gui-mao*) in the cycle of sixty. The start of the double-hour *chen* is 29 *ke* after midnight (to the nearest *ke*). Thus the solstice occurred 3 *ke* after the start of *chen* and its time and date are given as

gui-mao   chen   3 ke.

Cosmology

Besides straightforward astronomy, the Chinese also had speculative theories about the general structure of the universe. The first such theory to crystallize into a reasonably coherent whole was the *Gaitian* [Celestial lid] theory (Figure 5.4): that the heavens and the earth were parts of two spheres, the earth inside the heavens, though this theory maintained at the same time in some inexplicable way that the earth was square. Details in the *Zhou bei suan jing* [Orbit and gnomon arithmetic] about 100 B.C. seem to give the earth a radius of 225,000 *li* and the heavens a radius of 305,000 *li*. (One *li* is about half a kilometer.) This theory is thought to go back to legendary times. A second theory, *Huntian* [Celestial sphere], which goes back at least to 100 B.C., likens the heavens to an egg with the earth as its yolk. The heavens are half-filled with water and half with vapor; the earth floats on the water. By the end of the Han dynasty (A.D. 200) this was the generally accepted theory. The idea that the earth was round did not seem to percolate from these cosmologies into astronomy proper, nor into geography—Chinese maps were constructed as though the earth were flat. In another theory, *Xuanye* [Ubiquitous darkness], the

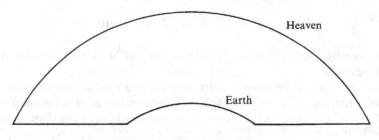

FIGURE 5.4. The *Gaitian* theory.

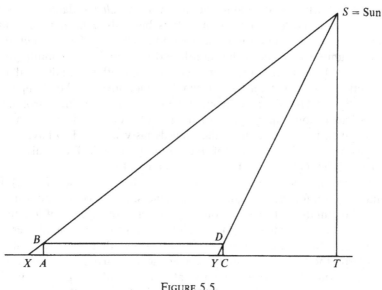

FIGURE 5.5.

sun, moon, and stars are composed of condensed vapor and float freely in space.

The Chinese also had an estimate of the height of the sun above the (flat) earth at noon on the day of the summer solstice. This was found by measuring the shadows of two vertical rods (perhaps like the ones described on page 27) a known distance apart, one due south of the other. In Figure 5.5, AB and CD are the rods, both the same length, and AX and CY are the shadows. By elementary geometry we can find the sun's height, ST in the figure, in terms of AB, XY, and the difference between the shadow-lengths.

If the rods are 8 *chi* high (a standard height) and 2,000 *li* apart, and if the shadows measure 1.5 and 1.7 *chi*, then the height of the sun is 80,000 *li*. Once we have found the height of the sun, we can, using the same geometrical technique, find the difference in shadow-lengths, for standard rods, in terms of their distance apart. If they are 1,000 *li* apart, the difference is $\frac{1}{10}$ *chi*, i.e., 1 *cun*. Consequently, the shadow of a standard rod shortens by one *cun* for every thousand *li* of southward travel. This figure is widely quoted in Chinese sources.

Almanacs

In ancient and medieval China each emperor would issue his own almanac. Unofficial astronomers were discouraged—they might be rebels preparing their own almanacs—and official astronomy was kept partly secret. The earliest known almanac, the *Xia xiao zheng*, dates from some time between 700 B.C. and 300 B.C. It is basically a farmers' calendar. Later almanacs contained more astronomy, including data for determining the number of days in the month and the number of months in the year, as well as instructions for sacrifices and other rituals, and even descriptions of the calamities that would befall anyone who skimped the rituals. By the Han dynasty (from 200 B.C. to A.D. 200) the astronomical details had become quite precise, though old ideas had not been completely given up: for example, the Han dynasty was said to have started 275 years after the capture of the *qi-lin* (a mythical animal), which occurred 2,760,000 years after the "beginning of time."

The best known almanac of that time, the *San tong* (7 B.C.) [64] contained data for the sun, moon, and planets; it also contained historical dates, the lengths of tubes for sounding certain notes, units of length and volume, and standards for temple buildings and for ceremonial dress.

Although Shih Shen (about 350 B.C.) knew that the moon moved irregularly, the Chinese did not discover the sun's irregularity until about A.D. 570. And they did not incorporate the moon's irregularity in their almanacs; as late as A.D. 200 they calculated the date of the new moon as though the motion were perfectly uniform.

They also included procedures for calculating the name (in the cycle of sixty names) of a given year and the name of the first day of the year, for finding whether the year has twelve or thirteen months, and for finding whether a given month has 29 or 30 days.

The rule for finding whether a year needs a thirteenth month is based on the discovery that 19 years is almost exactly equal to 235 months. Because 19 years contain an odd quarter of a day (on the basis of $365\frac{1}{4}$ days to the year) the Chinese introduced the *bu* of 4×19 years. The rule is as follows. Multiply the number of years elapsed in the *bu* by 235 and divide by 19. If the remainder is 12 or greater, the year needs a thirteenth month.

This rule makes 12-month and 13-month years occur in a fixed 19-year cycle

$$- + - - + - - + - + - - + - - + - - +$$

where − denotes a 12-month year and + a 13-month year. Although new almanacs were continually being introduced, this cycle remained in force throughout most of Chinese history. This 19-year, 235-month, cycle was probably derived from the Babylonians.

The Length of the Year

The length of the year is a basic figure needed for compiling an almanac. It was well known from early times to be just over 365 days, but various almanacs gave different figures for the odd fraction of a day over the 365. For example,

San tong almanac (7 B.C.)	385/1,539 days
Da ming almanac (A.D. 462)	9,859/39,491 days
Tian bao almanac (A.D. 550)	5,787/23,660 days
Da yen almanac (A.D. 724)	743/3,040 days

The obvious way to find the length of the year is to count the number of days between two summer solstices (or two winter solstices) a large number of years apart, and divide it by the number of years. This will give the average length of the year between these two dates. The Chinese, however, did not use this method. Here, for example, is how they obtained the figure for the *San tong* almanac [65]. First, they had a relation between the month and the day:

81 months contain 2,392 days.

They then appealed to the figure that they were using for deciding when a year needed a thirteenth month:

19 years contain 235 months.

It follows that the number of days in a year is

$$\frac{235}{19} \times \frac{2392}{81}$$

which comes to 365 + 385/1,539. It is this method that accounts for the rather fearsome fractions in the almanacs. The denominator of the final fraction (here 1,539) was called *ri fa* (literally "day denominator"). It seems to have been important: one source lists the *ri fa* for forty-seven almanacs [66].

Official Records

The twenty-eight official histories of the Chinese dynasties contain a wealth of astronomical information. The *Shi ji* [Historical records], which covers legendary and feudal times, describes the sun, moon, planets, meteors, oddly shaped clouds, comets, and their astrological significance. The *Qian Han Shu* [History of the early Han] adds a list of astronomical happenings and historical events supposed to be connected with them. The *Hou Han Shu* [History of the later Han] has three astronomical chapters mostly concerned with the connection between astronomical and historical events. The next, the *San guo zhi* [History of the three kingdoms], does not mention astronomy, but the fifth official history, that of the *Jin* dynasty (from A.D. 265 to 420), which was written by a committee of scholars, has three chapters on astronomy [67].

They start with a brief historical introduction, quoting from the *Yi jing* [Book of changes] "Heaven hangs up its symbols, from which are seen good and bad fortune; and the sage makes his interpretations accordingly." When the Emperor governs correctly there are no abnormalities such as eclipses. Next comes a description of the *Gaitian*, *Xuanye*, and *Huntian* cosmologies (see page 90): by later Han times *Xuanye* had died out and *Gaitian* was found defective; the *Huntian* theory is the only one that approximates to the truth, and the Imperial astronomer's instruments—presumably armillary spheres—are suited to it. Their method of use "will remain unchanged for a hundred generations." Some arguments against the *Gaitian* and *Xuanye* theories are mentioned: against the *Xuanye* theory Ge Hong argued: "If the stars and the *xiu* are not attached to the heavens, then the heavens serve no useful purpose, and so we can assert that they do not exist. What is the purpose of postulating heavens that remain stationary?"

Next comes a mention of the "circumpolar template"—a template which, when held up in the direction of the pole, marks the positions of the most prominent stars near the pole—and various armillary spheres, one of which was rotated by water power. There is a detailed description of the celestial sphere, including the position of the equinoctial points ($5\frac{1}{6}$ *du* in the first *xiu* and $14\frac{5}{16}$ *du* in the fifteenth *xiu*), and a detailed description of why daylight lasts longer in the summer. The diameter of the heavens is quoted from earlier books as

$$329{,}401 \ li \quad 122 \ bu \quad 2 \ chi \quad 2 \ cun \quad 1\tfrac{10}{71} \ fen$$

but so are calculations which lead to just half of this figure.

The next section is a long list of stars. A fairly typical extract is:

South of *Chuan she* in the Milky Way are the five stars of *Cao fu*, the officials in charge of horses. When these stars fail to appear, horses are scarce. Nine stars in the shape of a hook, known as *gou xing*, can be seen to the west in the Milky Way. A straight hook presages an earthquake.

Then comes a description of the course of the "River of Heaven" (the Milky Way). The chapter ends with a section on the twelve "Jupiter stations." The sidereal period of Jupiter is roughly twelve years, so we can divide the sky into twelve parts, in each of which Jupiter remains for about a year. Each of these is associated with one character from the sequence of twelve animals used in the sixty-name cycle, and corresponds to some part of the country. For example, the *Shou xing* station extends from the twelfth *du* of the twenty-eighth *xiu* to the fourth *du* of the third *xiu*, is associated with the character *chen*, and corresponds to what is now Shandong and Hebei.

Next we read about the "seven luminaries"—the sun, moon, and five planets. For example, in the presence of an infamous Emperor, or ministers who disrupt the state, the sun turns red and loses brightness. Also, an eclipse of the sun is caused by conflict between *yin* and *yang* and foretells that ministers will try to usurp the imperial power. The moon's speed varies, and the moon deviates from the ecliptic. If Mars is sickle-shaped and retrogresses, there will be a military defeat. Conjunctions of two luminaries also presage various happenings (mostly unpleasant ones).

Next come "miscellaneous celestial objects," classified into:

Rui xing	[auspicious stars]	Four are listed
Yao xing	[ominous stars]	Twenty-one are listed
Ke xing	[guest stars]	Sixteen are mentioned
Liu xing	[drifting stars]	Various types are mentioned
Rui qi	[auspicious vapors]	Three are listed
Yao qi	[ominous vapors]	Two are listed

"Auspicious stars" include the earth-shine on the moon. This category and the next three all include comets and meteors; *ke xing* includes also novas and supernovas and possibly variable stars, but none listed here can be identified. The "auspicious vapors" include auroras, and the "ominous vapors" include rainbows.

The next section deals with solar halos, and contains a large number of technical terms. For example, *mi*, which literally means "completion" and is described as a "white rainbow," is a parhelion.

Next comes a section on miscellaneous vapors (mostly clouds), and the chapter ends with a list of observed auroras, eclipses, halos, sun-spots, and conjunctions of two luminaries. For example,

On a *bing-xu* day in the tenth month of the first year of the reign of Ai Di the moon covered Venus in the tenth *xiu*. According to the standard prophecy, this meant disaster to Yangzhou. Loyang fell during the third year.

(*Bing-xu* is the twenty-third name in the cycle of sixty. The date is A.D. 363, November 26.)

The third and last astronomical chapter of the *Jin shu* continues with records of conjunctions, appearances, disappearances, and retrogressions of the planets, entrances of the moon and planets into the various *xiu*, and meteors. The records are extensive: an English translation occupies over 100 pages. They cover a period of 200 years, from A.D. 221 to A.D. 420.

Because oriental archives are so much fuller than Western ones, astronomers have combed them thoroughly for records of novae and supernovae. This is not as easy as it might seem, because the various kinds of "guest star" include very diverse kinds of spectacles in the sky; if a record mentions a prominent tail, or considerable movement across the sky, for example, it is probably a comet that was seen. According to F.R. Stephenson there are, in Chinese, Korean, and Japanese archives, about 75 possible novae and supernovae, one in 532 B.C., the others between 204 B.C. and A.D. 1604. Of these, only three (in A.D. 1006, 1572, and 1604) were noticed in the west [68].

Noon Shadows

Chinese almanacs do not list the observations on which their calculations are based. However, two particularly interesting sets of observations have been preserved, both concerned with measuring the length of the shadow of a vertical rod at noon (see page 41).

A Tang Dynasty Survey

The Chinese knew that the length of the shortest shadow at a particular place (at noon on the day of the summer solstice) varies from one place of observation to another, but they did not know the reason for this variation, namely that the earth is spherical; and they had a wrong belief about just how the shadow-length varies. For many years they thought that the change in the length of the shadow was strictly proportional to the north–south displacement; in fact, that the length decreased by 1 *cun* for every 1,000 *li* southward. It eventually became clear that this belief was incorrect, and in A.D. 724 astronomers carefully investigated mid-summer shadow-lengths of an 8-*chi* rod at four places in a north–south line, with the following results [69]:

Place	Distance between places	Shadow length
Huazhou		15.7 *cun*
	198 *li* 1,074 *chi*	
Biazhou		15.3 *cun*
	167 *li* 1,686 *chi*	
Xuzhou		14.4 *cun*
	160 *li* 660 *chi*	
Yuzhou		13.65 *cun*

Between the first and the last of these places the length of the shadow decreases by 2.05 *cun* over a distance of 526.9 *li*. The account of the survey, although it remarks that this result conflicts with the old belief, does not give a new improved *cun*-per-thousand-*li* figure. Instead it lists the midwinter and equinox noon shadow-lengths and the elevation of the pole at these four places. The figures show that these were calculated using 24 *du* for the obliquity of the ecliptic, a round-number value that the Chinese had established early on (24 *du* is about 23.66°). The elevations of the pole at the four places are:

Huazhou	35.3 *du*
Biazhou	34.8 *du*
Xuzhou	34.3 *du*
Yuzhou	33.8 *du*

Between the first and the last of these places there is a change of 1.5 *du* in 526.9 *li*, i.e., 351.3 *li* per *du*. The account of the survey uses this figure, together with the elevations of the pole at four quite widespread places, to find their distances from Yangcheng, the principal Chinese observatory of those days, where the elevation of the pole is 34.4 *du*. The results are:

Place	Pole Elevation	Distance from Yangcheng
Hengye	40 *du*	1,861 *li* 1,284 *chi*
Langzhou	29.5 *du*	1,826 *li* 1,176 *chi*
Jiaozhou	20.4 *du*	5,023 *li*
Linyi	17.4 *du*	6,112 *li*

These calculations imply, although the account does not specifically say so, that the Chinese astronomers have now given up the incorrect theory that change in *shadow length* is proportional to north–south distance and adopted the theory (which is correct) that change in *pole elevation* is proportional to north–south distance. We do not know what the new theory was based on, but it was not based on the results of this survey. If we compute the number of *li* required for a change of 1 *du* in pole elevation for the three individual links in the chain we do not get a constant result but three different figures:

Huazhou to Biazhou	397.2 *li* per *du*
Biazhou to Xuzhou	335.8 *li* per *du*
Xuzhou to Yuzhou	320.5 *li* per *du*

The places mentioned in the account are all identifiable. They lie reasonably well on a north–south line, their longitudes varying only from 106° to 115°. The line is a long one: Hengye is near the Great Wall, while Linyi is in Indochina.

The Exact Instant of the Solstice

One difficulty with using measurements at noon to find the solstice is that the instant of solstice, i.e., the instant when the sun is at its greatest or least declination, could be any time of day, not necessarily noon. Zu Chongzhi (A.D. 430–501) tackled the problem by measuring the shadow at noon on two consecutive days near the solstice, computing the change in shadow length in one day, and assuming that the theoretical noon shadow length always changed at this rate [70]. To illustrate the details it is convenient to use some measurements made in Beijing near the winter solstice of A.D. 1277. A few days before the solstice, the noon shadow was 79.4855 *chi*. The shadow lengthened after this, then shortened, and the first day after the solstice when the noon shadow was still longer than 79.4855 *chi* was 7 days after the first measurement. The figures are

noon on day 1:	79.4855 *chi*
noon on day 8:	79.541 *chi*
noon on day 9:	79.455 *chi*

At some instant between noon on day 8 and noon on day 9 the shadow theoretically must have been the same length as at noon on day 1, i.e., 0.0555 *chi* shorter than at noon on day 8. After the solstice the theoretical shadow shrinks by 0.086 *chi* per day, and at this rate the instant at which it is 0.0555 *chi* shorter than at noon on day 8 is 0.0555 ÷ 0.086 days (i.e., 0.6454 days) after noon on day 8. We now have two instants, namely noon on day 1 and the instant we have just calculated, when the shadow is theoretically the same length. The solstice is the instant halfway between them, namely 0.3227 days after the midnight separating days 4 and 5.

The calculation above is given in full in the Chinese source (the official history of the Yuan dynasty), the result being rounded off to 0.32. Although the calculations for the length of the shadow at the instant of solstice are not given, the same method must have been used, as the result quoted, 79.85 *chi*, is almost exactly what the method gives. (Actually, the Yuan history rounds the figure off to 79.8 *chi* but the history of the next dynasty, the Ming, quotes 79.85. My calculations give 79.86.)

Both these calculations are theoretically unsound because the noon shadow length does not increase at a constant rate. It is surprising that the Chinese did not realize this, as their data show it quite clearly. The records give a total of eighteen shadow lengths around the winter solstice of 1277 as follows [71]. (x 21 means the twenty-first day of the tenth month, etc.)

x 21	70.971	xi 22	79.455	xii 08	74.9595
xi 01	75.9865	xi 26	78.7935	xii 09	74.486
xi 02	76.377	xi 27	78.55	xii 12	71.9725
xi 09	78.6255	xi 28	78.3405	xii 15	71.343
xi 14	79.4855	xii 06	75.851	xii 16	70.76
xi 21	79.541	xii 07	75.417	xii 17	70.1565

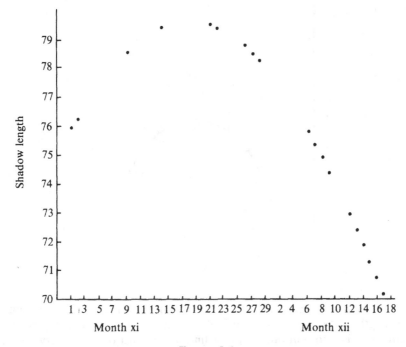

FIGURE 5.6.

xi 14 is the day that we called day 1 above, so that the solstice is in the afternoon of xi 17. It is clear that near this date the shadow length changes slowly, e.g., by 0.086 *chi* between xi 21 and xi 22, whereas further from the solstice it changes much faster, e.g., by 0.6035 *chi* between xii 16 and xii 17. The point is brought out quite clearly by a graph of shadow length against date (Figure 5.6). In fact, the rate of change is almost exactly proportional to the time interval from the solstice.

The effect of the Chinese theory is shown in Figure 5.7. The three spots are the three data used. The Chinese theory is that the slope of the line AX is the same as that of the line CBY and the solstice is given by the point Z where these lines cross. The correct solstice is given by Z^*, the highest point of the curve on which the data points lie. The time will be fairly accurate but the maximum shadow-length will be too long. The correct length is 79.66, rather than 79.86, *chi*.

These results attracted the attention of the noted French astronomer LaPlace, who made some comments on them that are not entirely justified [72]. He maintained that the Chinese used the interval-between-two solstices method—the one that I called "obvious"—rather than the method described on page 93. In fact, LaPlace said that an astronomer called Tsoutching compared observations he made at Nanking in A.D. 461

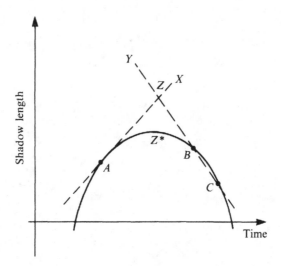

FIGURE 5.7.

with observations that had been made in Loyang in A.D. 173 and found the length of the year to be 365.24282 days. LaPlace did not give the source of his information. I cannot find any record of the observations, but it is possible to identify the astronomer. The *Da ming* almanac, promulgated in A.D. 462 was devised by Zu Chongzhi (French spelling, Tsou tch'ung-che) and gave the length of the year as 365 + 9,589/39,491 days. This equals 365.242815 days to six decimal places, so the name, date, and year length are consistent with LaPlace's statement. However, the *ri fa* 39,491 cannot possibly be produced by the method LaPlace described: the *ri fa* would have to be divisible by 288, the number of years between A.D. 173 and A.D. 461.

LaPlace was impressed by the accuracy of the figure 365.2428 compared with the estimate of the year length known in the West at that early date (Ptolemy's 365.2467 days. The correct value for the average length of the year at that date is 365.2423 days). However, it is rather misleading to compare the accuracy of a result obtained from a single calculation like this with other peoples' results. What matters is not the accuracy of a single result but the accuracy of the method. Suppose, for example, that each solstice can be found to the nearest fifth of a day. It would not be in the least remarkable if one solstice were out by 0.12 days and the other by 0.15 days. If the errors were in the same direction (both too early or both too late) the result would be out by 0.03/288 days, i.e., 0.00010 days; but if the errors were in the opposite directions the result would be out by 0.27/288 days, nine times as much. Generally, if the solstice can be found to the nearest n^{th} of a day and the two observations are m years apart, the accuracy that the method guarantees is $2/mn$ days. Later

Chinese estimates for the length of the year, reduced to decimals for easy comparison, were as follows:

A.D. 520	365.2437 days
A.D. 539	365.2442 days
A.D. 547	365.2243 days
A.D. 550	365.2446 days
A.D. 566	365.2443 days
A.D. 576	365.2445 days

The correct value of the average length of the year, to four decimal places, at these dates, according to modern astronomers, is

$$365.2423 \text{ days.}$$

Not until well after A.D. 1000 did Chinese estimates settle down to a value close to this. The *Tong dian* almanac of A.D. 1199 gave

$$365.2425 \text{ days.}$$

(By this time the correct value had come down to 365.2422 days: it decreases by about half a second per century. There are further details on page 125.)

Later Developments

Chinese astronomy reached its greatest heights in the Yuan dynasty, familiar to most of us as the dynasty during which Marco Polo visited China. The official history of this dynasty, the *Yuan shi*, contains two chapters on astronomy and six on calendrical science, written by Wang Xun, the court astronomer (*Tai shi ling*), and Guo Shoujing (Figure 5.8), his assistant, and finished by Guo after Wang died in 1282. It is in the chapters on astronomy that we find a detailed description of those instruments that so impressed an Italian visitor (see page 41), as well as some "western," i.e., Persian or Arab, instruments. We also find results of surveys, starting with a list of places with pole elevations 15 *du*, 25 *du*, 35 *du*, 45 *du*, 55 *du*, and 65 *du*, i.e., places evenly spread out in latitude from south of China to north of it (the first place is Nanhai, literally "south sea"; the last is Beihai, "north sea") to which the capital Dadu is added. For each place the length of the summer solstice shadow for an 8 *chi* rod and the lengths of day and night at the solstice are given. There follows a list of pole elevations for twenty important places, starting with Shandu (better known to us as Xanadu) and Beijing. Incidentally, most writers describe Dadu as an old name for Beijing, but the Yuan astronomers cannot have regarded them as the same place—they gave

FIGURE 5.8. Chinese stamp. The vertical line of Chinese characters reads: Guo Shoujing (A.D. 1231–1316), Yuan dynasty. (Collection E.C. Krupp, Griffith Observatory.)

them different pole elevations. The account of astronomy ends with extensive lists of eclipses, sunspots, conjunctions, and other astronomical phenomena connected with the sun, moon, and planets.

There is more information on astronomy in the calendrical chapters of the *Yuan shi*, the first four of which are devoted to the *Shou shi* almanac, which was promulgated in A.D. 1281. Here, for instance, is an extensive list of shadow measurements used for finding the solstice. I listed some of these (to be precise, all those from the winter of *Zhi Yuan* 14, i.e., A.D. 1277) on page 98 in chronological order, but in the *Yuan shi* they are rather more interestingly arranged. First, for finding the winter solstice of *Zhi Yuan* 14, come five sets of three (or in one case four) measurements, as shown in the left-hand column of Table 5.1. Figures like 14 xi 21 (43) are Chinese dates: this one is year 14 (of *Zhi Yuan*), month 11, day 21, and it is the forty-third in the sixty-name cycle. After each date comes the shadow length in *chi*.

TABLE 5.1

Year 14 winter solstice	Year 15 summer solstice
14 xi 14 (36) 79.4855	15 v 19 (38) 11.775
14 xi 21 (43) 79.541	15 v 28 (47) 11.78
14 xi 22 (44) 79.455	14 v 29 (48) 11.8055
(40) Fifth *shi* 3 *ke**; 4 days	(2) Twelfth *shi* 3 *ke*; 4 days
14 xi 09 (31) 78.6355	14 xii 15 (6) 71.343
14 xi 26 (48) 78.7935	15 xi 02 (18) 70.7595
14 xi 27 (49) 78.55	15 xi 03 (19) 71.406
14 xi 28 (50) 78.3045	156 days
8 or 9 days	14 xii 12 (3) 72.9725
14 xi 01 (23) 75.9865	14 xii 13 (4) 72.4545
14 xi 02 (24) 76.377	14 xii 14 (5) 71.909
14 xi 06 (57) 75.851	15 xi 04 (20) 71.9575
Fifth *shi* 3 *ke*; 17 days	15 xi 05 (21) 72.505
14 x 21 (13) 70.971	15 xi 06 (22) 73.0335
14 xii 16 (7) 70.76	158 or 9 days
14 xii 17 (8) 70.1565	14 xii 07 (58) 75.417
Fifth *shi* 3 *ke*; 27 days	14 xii 08 (59) 74.9595
14 vi 05 (60) 13.08	14 xii 09 (60) 74.486
15 v 01 (20) 13.0385	15 xi 09 (25) 74.5205
15 v 02 (21) 12.925	15 xi 10 (26) 75.0035
160 days	15 xi 11 (27) 75.4485
	163 or 4 days

There are similar data for the next three solstices.

* See page 90 for these units.

The first trio is the one used (page 97) to illustrate the Chinese method for finding the exact time of the solstice. The calculation is given in full in the *Yuan shi* (Figure 5.9); it shows that the solstice occurred 4 days 32 *ke* after the start of the day of the first observation. This result is presented as (40) fifth *shi* 3 *ke* (see page 90). It is followed, in a separate sentence, by the approximate result "4 days," which means that the solstice was on the fourth day after the date of the first of the three measurements. For the other four groups of measurements the calculations are not given, only the results, and sometimes only the approximate result; these agree with the result from the first group and presumably function merely as a check. The dates of the last group are surprisingly far apart and are not even in the winter of *Zhi Yuan* 14, but in the preceding and following summers.

Similar remarks apply to measurements and calculations for the other solstices, except that the results for the last of them do not agree completely but range over four consecutive days.

I have given the shadow-length data in full because they are a particularly good example of precise and comprehensive Chinese measurements,

推至元十四年丁丑歲冬至
其年十一月十四日己亥，景長七丈九尺四寸八分五釐五毫；至二十一日丙午，景長七
丈九尺五寸四分一釐；二十二日丁未，景長七丈九尺四寸五分五釐。以己亥、丁未二日之
景相校，餘三分五（釐）〔毫〕爲晷差，（三）進二位，以丙午、丁未二日之景相校，餘八分六釐爲
法，除之，得三十五刻；用減相距日八百刻，餘七百六十五刻；折取其中，加半日刻，共爲四
百三十二刻半；百約爲日，得四日；餘以十二乘之，百約爲時，得三時；滿五十又作一時，共
得四時；餘以十二收之，得三刻；命初起距日己亥算外，得癸卯日辰初三刻爲丁丑歲冬至。
此取至前後四日景。
十一月初九日甲午，景七丈八尺六寸三分五釐五毫；至二十六日辛亥，景七丈八尺七
寸九分三釐五毫；二十七日壬子，景七丈八尺五寸五分。以甲午、壬子景相減，復以辛亥、
壬子景相減，準前法求之，亦得癸卯日辰初三刻。至二十八日癸丑，景七丈八尺三寸四釐
五毫，用壬子、癸丑二日之景與甲午景，準前法求之，亦合。此取至前後八九日景。
十一月丙戌朔，景七丈五尺九寸八分六釐五毫；二日丁亥，景七丈六尺三寸七分七釐；
至十二月初六日庚申，景七丈五尺八寸五分一釐。準前法求之，亦在辰初三刻。此取至前
後一十七日景。
十（二）月二十一日丙子，（四）景七丈九寸七分一釐；至十二月十六日庚午，景七丈七寸
六分；十七日辛未，景七丈一寸五分六釐五毫。準前法求之，亦得辰初三刻。此取至前後
二十七日景。
六月初五日癸亥，景一丈三尺八分；距十五年五月癸未朔，景一丈三尺三分八釐五毫；
初二日甲申，景一丈二尺九寸二分五毫。準前法求之，亦合。此取至前後一百六十日景。
推十五年戊寅歲夏至
五月十九日辛丑，景一丈一尺七寸七分七釐五毫；距二十八日庚戌，景一丈一尺七寸
八分；二十九日辛亥，景一丈一尺八寸五釐五毫。用辛丑、庚戌二日之景相減，餘二釐五
毫，進二位爲實；復用庚戌、辛亥景相減，餘二分五釐五毫爲法；除之，得九刻；用減相距日
九百刻，餘八百九十一刻；半之，加半日刻，百約，得四日；餘以十二乘之，百約，得十一時；

志第四 曆一 一一二三

FIGURE 5.9. The start of the shadow-length calculations in the Yuan Shi (see page 103). The columns are read from top to bottom, the right-hand column first. With a knowledge of the Chinese numbers

一	二	三	四	五	六	七	八	九
1	2	3	4	5	6	7	8	9

it is easy to read the data in the text.

from a time when astronomy in the Western world was in the doldrums. They acquired considerable prestige in Europe when the nineteenth-century French astronomer LaPlace admired their precision and used them to confirm that the angle between the ecliptic and the equator had decreased between their time and his.

A little later in the *Yuan shi* comes a list of forty-eight winter solstices spread through Chinese history comparing the "observed" dates (presumably calculated from shadow lengths) given to the nearest day with the dates computed according to six different almanacs and quoted to the nearest hundredth of a day. The sixth of the almanacs is the *Shou-shi*, and the dates computed from it agree with the observed dates over the whole range. (According to modern calculations the *Shou-shi* almanac was out by about three days for dates around 700 B.C., by about a tenth of a day around A.D. 700, and by about three-hundredths of a day around A.D. 1000; it was correct to the nearest hundredth of a day from 1250 onward [73].)

The rest of the material in these chapters is more typical of what one would expect to find in an astronomical almanac: figures for the length of the year and the month, the positions of the solstices on the celestial globe, the positions of the *xiu*, tables from which the positions of the sun and moon on any date can be found, similar tables for the points where the moon's path crosses the sun's, leading to the computation of the times and magnitudes of eclipses, and a list of all eclipses observed since records started.

The *Shou-shi* calendar is based on two remarkably accurate figures: estimates of the length of the year and of the obliquity of the ecliptic. Both are worth looking at in some detail.

The Length of the Year

We have seen (page 93) how the older Chinese almanacs computed the length of the year as a complicated fraction, whose denominator is called a *ri fa*. With the *Shou-shi* almanac comes a change; although the *Yuan shi* lists the *ri fa* for forty-one earlier calendars, it states that the *Shou-shi* almanac does not use a *ri fa*, and quotes the length of the year not as a fraction but as a decimal: 365 days 24 *fen* 25 *miao*, i.e., 365.2425 days. This is within 20 seconds of the correct figure, 365.2423. The *Yuan shi* does not say how this result was obtained. Modern Chinese commentators have suggested that it is the result of dividing 4,382,910 by 12,000, both figures coming from an earlier almanac, the *Tong tian* of A.D. 1199, and indeed 12,000 is listed in the *Yuan shi* as the *ri fa* for the almanac [74]. But not only would this contradict the statement in the *Yuan shi* that the *Shou-shi* almanac did not use a *ri fa*; it leaves completely open the question of where the earlier calendar obtained these figures. In fact, all that we can really say is that we do not know how the Chinese arrived at their accurate result.

The Obliquity of the Ecliptic

The *Shou shi* calendar uses the value 23.903 *du* for the angle between the celestial equator and the ecliptic, i.e., the obliquity of the ecliptic. (The figure is found in a table for the declination of the sun in terms of its distance from the solstice: when this distance is zero the declination is given as 23.903 *du* [75].)

23.903 *du* is equal to 23°33′34″ to the nearest second and is remarkably accurate, the correct figure at that date according to modern calculations being 23°32′01″, which equals 23.877 *du*. The *Yuan shi* does not say how the result was obtained, but the standard method (as described on page 44) is to use the elevation of the sun at the two solstices. If we use the Chinese figures of 79.85 and 11.71 *chi* for the shadow lengths of a 40-*chi* rod, and the Chinese method of interpolation, we obtain 23.88 *du*. If we interpolate correctly in the Chinese shadow-length data we obtain 23.84 *du* for the obliquity. It is ironic that the incorrect interpolation turns this good result into an even better one.

Note. Some modern commentators appear to think that the figure for the obliquity was obtained from the following two angles, which are given in the *Yuan shi*:

> angle between pole and summer solstice   67.4113 *du*,
> angle between pole and winter solstice   115.2173 *du*.

These two figures are inconsistent with each other: for the first to be correct the obliquity would have to be 23.9012 *du*; for the second to be correct it would have to be 23.9048 *du*. Moreover, these angles cannot be basic data; no angles could be measured with the precision quoted and they must be the results of calculation.

Celestial Motions

Parts of the chapters on the *Shou-shi* almanac are based on very sophisticated mathematics, presumably developed for this purpose. Some of the calculations, for example those which tell how much progress along the ecliptic corresponds to a given angular progress along the equator, are equivalent to the use of spherical trigonometry; and a table giving the irregular progress of the moon along its orbit uses a formula of degree three. (This makes it all the more puzzling that the Chinese used only linear interpolation, that is, interpolation of degree one, in the shadow-length calculations.)

Let us now look at the table for the motion of the moon [76] (opposite). The twenty-seven columns refer to the twenty-seven days in an anomalistic period of the moon; they are numbered in the top row. The second row gives the successive values of a zig-zag function (see page 76) with a daily difference of 12.20. It is easy to calculate that this function goes through a complete cycle in exactly $27\frac{33}{61}$ days, so this is the length of the anomalistic period. The third row is an intermediate calculation; it is obtained from the second row by applying a formula of degree three. The fourth row gives the number of *du* covered in each day as the moon progresses round its orbit. The last row gives the total motion from the beginning of the period, and is obtained by summing the daily motions.

The intermediate calculation is explained in the *Yuan shi* as follows. To set up the fast/slow table, first multiply the time in days by 12.20 and if this is greater than 84 subtract it from 168. The result is the *mo xian*

0	1	2	3	4	5	6
0	12.20	24.40	36.60	48.80	61	73.20
0	1.3077	2.4963	3.5305	4.3748	4.9938	5.3522
14.6764	14.5573	14.4029	14.2130	13.9877	13.7271	13.4446
0	14.6764	29.2337	43.6366	57.8496	71.8373	85.5644

7	8	9	10	11	12	13
82.60	70.40	58.20	46	33.80	21.60	9.40
5.4281	5.2947	4.8735	4.1996	3.3086	2.2359	1.0168
13.2353	12.9475	12.6948	12.4777	12.2960	12.1496	12.0462
99.0009	112.3443	125.1918	137.8866	150.3643	162.6603	174.8099

14	15	16	17	18	19	20
−2.80	−15	−27.20	−39.40	−51.60	−63.80	−76
−0.3088	−1.5923	−2.7488	−3.7422	−4.5380	−5.1004	−5.3938
12.0852	12.2122	12.3752	12.5730	12.8063	13.0753	13.3377
186.8561	198.9413	211.1535	223.5287	236.1017	248.9080	261.9833

21	22	23	24	25	26	27
−79.80	−67.60	−55.40	−43.20	−31	−18.80	−6.60
−5.4248	−5.2223	−4.7399	−4.0131	−3.0772	−1.9677	−0.7201
13.5712	13.8511	14.0955	14.3046	14.4782	14.6163	14.7154
275.3210	288.8922	302.7433	316.8388	331.1434	345.6212	360.2379

(literally end limit); it is a zig-zag function of time and is given in the second line of the table. Multiply the *mo xian* by 325 and add 28,100. Multiply the result by the *mo xian* and subtract 11,110,000. If we denote the *mo xian* by x, the final result is $x[x(325x + 28,100) − 11,110,000]$, which is a third-degree expression in x. It is given in the third line of the table.

The same effect is obtained in a different way in the *Da tong* calendar used in the next (Ming) dynasty and explained in the official history (*Ming shi*, Chapters 31–36). A table for the motion of the sun (*Ming shi*, Chapter 33, page 624) starts as follows:

0	1	2	3
4.9386	4.9572	4.9758	4.9944
510.8569	505.9183	500.9611	495.9853
0	510.8569	1,016.7752	1,517.7363

and has 89 entries, covering the 89 days from the winter solstice halfway to the following summer solstice. (The Ming astronomers knew the length of time from one solstice to the next: 178 days from winter to summer, 187 days from summer to winter. In fact, they had more precise figures: 177.82 and 187.42 days.) The bottom row gives the difference between the distance actually traversed by the sun and the distance it would traverse if it traveled at a constant speed throughout the year. The unit is a ten-thousandth of a *du*. The top two rows are for calculation. The top row starts with 4.936 and the entries increase by 0.0186 each time. The second row starts with 510.8569 and each successive entry is obtained by subtracting the entry above; for example, the 505.9183 is 510.8569 − 4.9386. The bottom row starts at zero and the successive entries are obtained by adding successive entries from the second row.

It is clear that the entries in the second row are the differences between successive entries in the bottom row: let us call them "first differences." Similarly, the entries in the top row are the differences between successive first differences: let us call them "second differences." Finally, the difference between successive second differences is 0.0186 each time: the third difference is constant.

It is a mathematical fact that a table constructed in this way from a constant third difference is equivalent to a third-degree formula. We can, if we like, compute the formula corresponding to this particular table: the entry under day x will be

$$x[513.32 - x(2.46 + 0.0031x)].$$

The next table covers the 94 days from the summer solstice halfway to the following winter solstice. It is built up similarly from a constant third difference 0.0162. These two tables between them, using symmetry, cover the whole year.

The tables for the motion of the planets are also interesting. I reproduce here the table for Jupiter [77]. By comparing it with the table on page 86 we can see how much progress was made in 1,200 years.

	ji chu	ji mo	chi chu	chi mo	liu	tui
16.86 days	28	28	28	28	24	46.58
3.86 *du*	6.11	5.51	4.31	1.91		$4.8812\frac{1}{2}$
2.93 *du*	4.64	4.19	3.28	1.45		$0*.3287\frac{1}{2}$
23 *fen*	22	21	18	12		

	tui	liu	chi chu	chi mo	ji chu	ji mo
46.58	24	28	28	28	28	16.86
$4.8812\frac{1}{2}$		1.91	4.31	5.51	6.11	3.86
$0*.3287\frac{1}{2}$		1.45	3.28	4.19	4.64	2.93
16			12	18	21	22

The table starts at conjunction. The next four columns give the progressive motion when Jupiter is a morning star, the following four give the retrograde motion at opposition and the four after them the remaining progressive motion. Presumably, the last column gives half the period of invisibility (the first column giving the other half). Each period of progressive motion is divided into an accelerating (*chi*) and a decelerating (*ji*) phase; each phase is divided into an initial (*chu*) and final (*mo*) half. The figures in the second row give the length of each segment of the motion in days, and the next two rows give the number of *du* covered, the third row giving mean movement and the fourth row true movement. The last row gives the daily movement (a *fen* is a hundredth of a *du*). The character that I have denoted by 0^* is not the usual zero symbol (a circle) but the character *kong*, meaning "empty."

The End of the Story

Independent Chinese astronomy came to an end when China was opened up to Western missionary influence at the end of the sixteenth century. Although the astronomer–priests brought to China some of the advances that had been made in Europe, such as a knowledge of the size and shape of the earth, and tables of positions of the planets and the moon far more accurate than any the Chinese had, these astronomers, being Jesuits, had to teach the Ptolemaic system. Ironically, therefore, some of the Chinese cosmological beliefs on which the Jesuits poured scorn were more correct than those held by the Jesuits. In 1595 one of them, Matteo Ricci, wrote that the Chinese thought that there was only one celestial sphere, not ten; they thought that the stars move in the void, instead of being attached to the firmament; and they thought that there was a vacuum in outer space where Westerners believed that there was air [78].

Chapter 6

The Greeks

The Early Thinkers

Starting some time before 700 B.C., yet another group of people, the classical Greeks, were coming to grips with astronomy.

Greek intellectual development is so important in the history of Western civilization that historians of science are all too often disappointed when they discover how little of what is generally said about the early Greek thinkers is known to be true with any certainty. Every reference—first-, second-, third-hand or yet more remote—has been investigated, expounded upon, and widely repeated; often too widely.

Greek scientific thought and speculation started with a school of philosophers in Ionia in the sixth century B.C. Prominent among them was Thales. The historian Herodotus (fifth century B.C.) tells us that Thales foretold an eclipse of the sun, which occurred during a battle between the Lydians and the Persians [79]. What Thales predicted, according to Herodotus, was the year in which the eclipse would take place; and that is usually thought to be nonsense. But not by everybody. Willy Hartner has given an explanation of how Thales might have foreseen the eclipse and an ingenious, though far-fetched, suggestion why he announced only the year [80]. (Hartner suggests that Thales knew of a tendency of eclipses to repeat after 47 years, and knew of an eclipse 47 years earlier.)

Thales suggested that the earth floats on water, whereas his near-contemporary, Anaximander, said that the earth is necessarily at rest because of its *homoiotes* [uniformity] and so does not need to rest on anything.

Heraclitus (540–480 B.C.) seemed to be coming to realize that the universe behaves in a periodic fashion; he also thought that the sun is a foot wide and new every day. Parmenides (512–400 B.C.) thought that the universe is spherical and realized that Hesperus and Eosphoros are the same planet, and that moonshine is reflected sunlight. Anaxagoras (500–428 B.C.) suggested that *nous* [mind] controls the universe; that comets are formed by planets colliding; that the sun is a fiery ball bigger than the

Peleponnesus (an improvement on Heraclitus); that the earth is flat, solid, and supported on air; that there are invisible bodies beneath the stars; that the moon is closer to the earth than the sun is; that eclipses of the moon are caused by the shadow of the earth and other bodies; and that eclipses of the sun are caused by the moon.

The followers of Pythagoras (sixth century B.C.) thought that the center of the universe is occupied by a fire which powers the moving heavenly bodies; that the earth moves round the fire, screened from it by a body called *antikhthon* [counter-earth]; that outside the earth's orbit are those of the moon, the sun, the planets, and the stars; that celestial motion produces the "music of the spheres," which we do not notice because it has always been there and we do not know what its absence is like; and (probably) that the earth is a sphere.

It would be easy to sneer at some of these ideas as naive, but there is a steady progress from the beginning of this period to the end. Had these thinkers not been inquisitive and ready to speculate (could the earth be moving? no one else even considered the possibility) how would later theories ever have started?

The Classical Greeks

Meton and Euctemon

Just after 450 B.C. the Greeks began to write astronomical and meteorological diaries called *parapegmata* (and continued to write them for at least three hundred years) [81]. The early *parapegmata* mention two astronomers by name, Meton and Euctemon. We have only tantalizing fragments about these two pioneers. They made observations in Thrace, Macedonia, the Cyclades islands, and Athens; and their pupils determined the date of the summer solstice in 432 B.C. [82]. They used the signs of the zodiac to describe positions on the ecliptic. Meton suggested that a period of 19 years, which contains almost exactly a whole number (235) of months, could be used to correlate the solar and lunar calendars [83].

Euctemon gave the lengths of the seasons as 90, 90, 92, and 93 days. This is a very significant fact. First, Greek astronomy is now getting down to measuring and counting. Second, the Greeks are beginning to notice the small irregularities that pervade astronomy and are so important in all but the most superficial investigations.

What exactly did Euctemon say? He gave the times between solstices and equinoxes:

From summer solstice to autumn equinox: 90 days
From autumn equinox to winter solstice: 90 days
From winter solstice to spring equinox: 92 days
From spring equinox to summer solstice: 93 days

The idea of equinox is quite simple. Nights are long in winter, days are long in summer; somewhere in between they will be equal. But not exactly halfway in between: the solstices alone show that the sun does not move at a constant speed round the ecliptic. If it did, the time from summer solstice to winter solstice would be equal to the time from winter solstice to summer solstice, and Euctemon made these times 180 days and 185 days, respectively. How could he find the date of equinox? Not directly, by timing the length of daylight, because refraction makes the sun seem to be on the horizon when it is really below it. A better way is to find when the sun sets due west, though this, of course, will give the equinox only to the nearest day. Later, about 150 B.C., Hipparchus invented an ingenious device for finding the equinox quite accurately. He set up a thin metal ring parallel to the plane of the equator. When the shadow of one half of the ring falls on the other half, the sun is on the equator and it is the time of the equinox.

Euctemon's observations do not seem to have been very accurate. Callipus, about 330 B.C., found 92, 89, 90, 94 days, respectively, which are correct to the nearest day.

The Greek Zodiac

Once the equinox points had been found, the Greeks had an alternative way of placing the signs of the zodiac on the ecliptic, using the equinoxes instead of the stars as reference points. From the time of Eudoxus (about 380 B.C.) onward, they made the zodiac start at the spring equinox. They named the signs after the constellations near them. (See page 67 for the Greek names for the signs.) At that date the spring equinox was in the constellation that we call by its Latin name Aries, so the first sign is called Aries. As time passed, the constellations moved relative to the equinox point, and soon the equinox point, though still called "the first point of Aries," found itself in the constellation Pisces. It is now about to enter Aquarius ("the age of Aquarius").

Eudoxus

Greek astronomy reached a new level of sophistication with Eudoxus (who was born between 408 and 390 B.C., and died at the age of 53), a mathematician of genius, being the inventor of the powerful technique known later as the "method of exhaustion" and of the treatment of incommensurable quantities in the fifth book of Euclid [84]. He found a way of describing the motions of the heavenly bodies by means of spheres. As long as we cannot measure the distances of the stars, we might as well assume that they are all the same distance from us, and so lie on a (large) sphere at whose center we are situated. The same applies to, say, the moon: so long as we are not concerned with (or cannot detect) any change in its distance from us, we might as well regard it as moving on a sphere. This is what Eudoxus did: in his theory the stars were fixed to a sphere, and the sun, moon, Mercury, Venus, Mars, Jupiter, and Saturn were each fixed to a sphere. The sphere of the stars rotated at constant speed; Eudoxus's feat was to devise an ingenious linkage of spheres which reproduced some salient features of the motion of the planets.

Was Eudoxus justified, in the light of what was known at the time, in assuming that each celestial body remained at a constant distance from the earth? For the sun: *yes*. We moderns know that the distance between the earth and the sun varies from 147 to 152 million kilometers, but this difference is indetectable: even the distance itself is hard to measure. Neither Copernicus (A.D. 1450) nor Kepler (A.D. 1600) had any idea of its true value, and the first reasonably accurate measurement was by Cassini in 1673. For the moon, the answer is *no*. The moon shows a distinct disk, whose apparent size varies with the distance of the moon from the earth. The ratio of the greatest apparent diameter to the smallest is 1.14 to 1, and this difference is detectable. For the outer planets, the answer is also *no*. Mars, for example, is much nearer the earth at some times than at others, and although the disk of Mars is too small for its apparent change in size to be detected by the naked eye, Mars seems much brighter when it is near than when it is far away: up to twenty-five times as bright. For Mercury and Venus, the answer is more difficult because both when they are at their closest and when they are at their farthest they are too near the sun to be seen. For the stars the answer is *yes*. To detect the difference between the distances of the various stars was utterly beyond early astronomers. Not until A.D. 1838 did anyone measure the distance of a star. Admittedly in A.D. 70 Geminus wrote:

> We must not suppose that all the stars lie on one surface, but rather that some of them are higher and some lower. It is only because our sight can only reach to a constant distance that the difference in height is imperceptible to us [85].

But Geminus did not convince anybody.

Nothing that Eudoxus wrote survives, and we know of his theories only second-hand. The spheres we mentioned were described by Aristotle (384–322 B.C.) and in more detail by Simplicius (sixth century A.D.) who wrote extensive commentaries on Aristotle's writings [86].

Before we look at the system in detail we must explain a technical term. The line through the center of a sphere at right angles to a great circle will meet the sphere in two points. These are called the "poles" of the circle. (Thus the poles of the celestial equator are the north and south celestial poles.)

Let us start with the moon. If we neglect the moon's deviation from the ecliptic (i.e., its latitude) we can represent its motion as follows. Think of the ecliptic as a great circle drawn on the celestial sphere, and think of the moon as crawling like a spider round the ecliptic, going completely round in a certain length of time that we call the sidereal period of the moon. To get this effect, Eudoxus imagined a sphere a trifle smaller than the celestial sphere pivoted to it at the poles A and B of the ecliptic, and turning on AB as axis at one revolution per sidereal period. If the moon is fastened to this second sphere at a point equally distant from A and B it will move round the ecliptic at the correct rate. However, Eudoxus knew that the moon deviated a little from the ecliptic, and that its greatest deviation did not always take place at the same position on the ecliptic. To account for this, he postulated a third sphere, carried inside the second. The angle between the axes of the second and third spheres equals the greatest angle by which the moon deviates from the ecliptic. The period of rotation of the third sphere is presumably the time actually taken for the position of greatest deviation to make a complete circuit of the ecliptic (about 18.61 years).

There is a serious flaw in this picture. The motion of the spheres will make the moon's latitude change very slowly. It will take over 18 *years* from the greatest northward latitude back to the greatest northward latitude again, whereas in actual fact this time is about 27 *days*. Perhaps Simplicius mistook the period of the second sphere for the period of the third sphere and vice versa; and, indeed, if the second sphere rotates once in 18.61 years and the third sphere rotates once in a latitudinal period (which is very nearly the same as a sidereal period), the motion of the moon in latitude will be correct. Another flaw in Eudoxus's system is that the spheres make the moon move at constant speed round the ecliptic, whereas its speed actually varies (and the Babylonians, at least, knew this).

Eudoxus's construction for the sun is similar, but its flaws are more serious. For one thing, the sun does not deviate from the ecliptic, and so the third sphere should not be there. And it is a bad mistake to have the sun move at a constant speed because it was quite well known at this date that it did not do so: Meton and Euctemon had noted the inequality of the seasons.

The main fascination of Eudoxus's system lies in the ingenious way in which it dealt with the planets. A planet moves very irregularly round the ecliptic. The only irregularity that Simplicius mentioned is the deviation in latitude, which is quite small (up to 7° for Mercury, not more than 5° for the others, and only about 1° for Jupiter) but there can scarcely be any doubt that the main phenomenon that Eudoxus wanted to reproduce is retrogression. Eudoxus gave each planet four spheres. The two outer spheres are just like those for the sun and the moon; they produce regular motion round the ecliptic in the sidereal period of the planet.

Let us consider the third and fourth spheres on their own (see Figure 6.1). Imagine a sphere $\mathscr{S}$ pivoted so that it can turn about a vertical axis AB. Inside it another sphere $\mathscr{T}$ is pivoted so as to turn about an axis PQ,

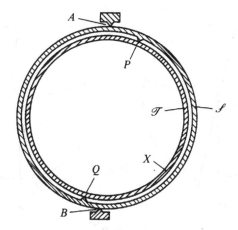

FIGURE 6.1.

where P and Q are points fixed on $\mathscr{S}$, P being near to A (and therefore Q near to B). $\mathscr{S}$ is set rotating, and $\mathscr{T}$ is set rotating relative to $\mathscr{S}$ at the same rate in the opposite direction. We might imagine a small meccano motor bolted to the floor at B, driving $\mathscr{S}$ clockwise at, say, one revolution per minute; and another motor bolted to $\mathscr{S}$ at Q, driving $\mathscr{T}$ anticlockwise at one revolution per minute. X is a spot of paint on $\mathscr{T}$ equally distant from P and Q; that is, on the great circle whose poles are P and Q. How does X move?

If PQ were vertical, the two rotations would cancel and $\mathscr{T}$ would remain as still as a man walking up a down escalator at precisely the speed of the escalator. If, however, PQ is not quite vertical, the rotations will not quite cancel. Figure 6.2(a, b), shows that X will move from an angle α below the horizontal to the same angle above the horizontal, once in each revolution: here α is the angle between PQ and AB. X does not, however, move up and down a vertical arc: the horizontal effects of the two rotations do not quite cancel. The precise path of X can be found mathematically: it is a figure-of-eight curve called a *hippopede* (see Figure 6.3). Neugebauer has shown how it could have been found, using only the geometry known in Eudoxus's time [87].

If we imagine the point X to be a planet, the effect of the spheres $\mathscr{S}$ and $\mathscr{T}$ (Eudoxus's third and fourth spheres) is to make the planet move round the hippopede like a toy train round a figure-of-eight track. If we now fit these spheres inside the second sphere, the hippopede will move round the ecliptic. Then the motion of the planet relative to the sky is as follows: it moves round an imaginary hippopede while the hippopede moves round the ecliptic (see Figure 6.4). The breadth of the hippopede carries the planet off the ecliptic, first to one side and then to the other. And if the speed of the planet in the hippopede is greater than the speed

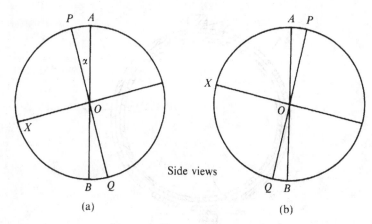

Side views

(a)   (b)

FIGURE 6.2.

of the hippopede along the ecliptic, then while the planet is moving back along the hippopede it is retrogressing.

So far, the construction is only qualitative. Can it be made quantitative— could Eudoxus produce the right amount of deviation by using the right speeds of rotation and judiciously choosing the angle between the axes of the third and fourth spheres? Simplicius quoted sidereal and synodic periods (all reasonably accurate except for the synodic period of Mars, which is about three times the correct value). The angle between the axes AB and PQ will determine the length and breadth of the hippopede. If it is chosen to give the right breadth (so as to give the correct value for the greatest deviation in latitude) it may or may not give the right length (which is required to give the right value for the "retrograde arc"—the amount of the ecliptic covered during a retrogression). Schiaparelli investigated this, and found that the system does not work for the outer

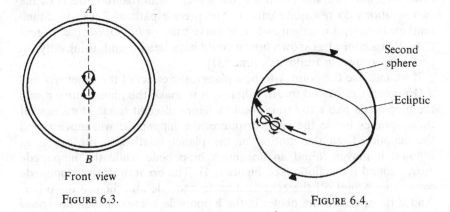

Front view

FIGURE 6.3.   FIGURE 6.4.

planets. However, for Venus and Mercury the results are not too bad.
[88].

In any case, the theory does not represent the motions very well: it
gives the wrong proportion of the synodic period spent in retrogression
and it makes the planets cross the ecliptic four times in each synodic
period; in fact, they cross it only twice.

Callipus (who lived about 370–300 B.C.) tried to improve the system
[89]. He used the same arrangement as Eudoxus for Saturn and Jupiter,
but inserted two extra spheres each for the sun and moon, and one extra
for each of the other planets. No explanation by Callipus himself of the
reason for the extra spheres is known, but Eudemus (about 330 B.C.) gave
one: that three spheres are not enough to account for the inequality
of the seasons discovered by Meton and Euctemon. Eudemus also explained why Callipus had only one extra sphere each for Mars, Venus,
and Mercury, but, unfortunately, the explanation is lost.

Besides *Peri takhon*, the book in which he described these rotating
spheres, Eudoxus wrote two other astronomical works, now lost:
Phaenomena and *Enoptron* [Mirror]. In the third century B.C., Aratus
wrote a poem, entitled *Phaenomena*, based on Eudoxus's work, and
Hipparchus wrote a commentary on this poem.

A typical quotation from the poem is:

Below the head (of the Great Bear) is Gemini; below her waist is Cancer; below
her hind legs Leo shines brightly [90].

Much of Hipparchus's commentary consists of detailed corrections of
Aratus's descriptions: for example, the head of Ophiuchus does not lie on
the tropic of Cancer as Aratus said, but 7° south of it.

There is a rather odd papyrus, probably written about 190 B.C., called
"*Techne Eudoxou*" [Art of Eudoxus] because it contains an acrostic of
twelve lines whose initial letters spell out this title (ch is one letter, X,
in Greek). The acrostic also contains 365 letters. The contents of the
papyrus are of no interest and its connection with Eudoxus is purely
nominal.

Aristotle

We come next to one of the giants of Greek philosophy: Aristotle (384–
322 B.C.). His treatise *De caelo* is astronomical philosophy rather than
astronomy, but in one passage in *Metaphysics* he did deal with real
astronomical details: he described Callipus's modification to Eudoxus's
system, with a further modification of his own.

As far as we know, Eudoxus and Callipus thought of their system as
merely a mathematical construction, that is to say, as a system for combining spherical motions; the spheres did not actually exist—just as the
modern theory of gravitation does not postulate invisible cords stretch-

ing from the earth to the moon. Aristotle, however, supposed that the spheres were real. This caused a complication. He could articulate the spheres for the outermost planet, Saturn, to the celestial sphere, but when he tried to articulate the spheres for Jupiter to the celestial sphere, the spheres for Saturn were in the way. Therefore, he had to add "unrolling" spheres, "one fewer than the spheres which combine to produce the planet's motion." Similarly for the other planets.

Aristotle gave several reasons for believing the earth to be spherical. First, since the natural movement of the element earth is downward toward the center, the particles of earth will converge into a sphere. Second, the edge of the dark part of the moon in an eclipse is always convex, as opposed to the edge of the dark part in the ordinary succession of the phases of the moon, which can be convex, concave, or straight. Third, certain stars visible from Egypt are not visible from further north, and certain stars which rise and set when seen from Egypt remain above the horizon to viewers further north; the changes are so great that the earth cannot be large.

Aristotle had to consider whether or not the earth moves—since the time of the Pythagoreans, no philosopher could ignore this point. He noted that heavy bodies fall toward the center of the earth (which is the center of the universe) because it is their natural tendency. This would therefore be the natural tendency of the earth if it were not at the center; so it must be at the center, held there by this tendency. It is not, as Thales thought, floating on water, nor, as Anaximenes thought, floating on air.

The stars according to Aristotle are not hot themselves; their light and heat come from friction with the air. They are too far off to warm us, and the moon is too slow; only the sun is both near enough and fast enough to do so.

The Size of the Earth

The earth's size and shape are vital to astronomy. Once we have realized that the earth is round, one astronomical phenomenon becomes less surprising: the fact that the pole round which the stars revolve is not directly overhead. On a spherical earth, "overhead" varies in direction from place to place, and it would be a coincidence if the pole were overhead to *us*.

The first inkling that the earth is not flat probably came from seafarers. The irregularities of the land mask any sign of its curvature, but on a calm day at sea the curvature shows quite clearly: a ship a long way away seems mysteriously below sea-level. Sailors call this "being hull-down on the horizon." Since this happens equally in all directions, the earth is a sphere, not a cylinder. (This argument does not appear in writing earlier than Strabo, about 10 B.C.; however, Strabo says that Homer knew of it.)

Strictly speaking, only the part of the earth which had been explored was seen to be spherical, but no one seems to have wondered whether the remote parts were of a different shape. We know today that the earth is not quite spherical: the diameter from pole to pole is 42 km less than the diameter across the equator, but 42 is not much in 6,500. To describe the earth as "tangerine-shaped" is rather misleading: it is far less flattened than any tangerine that ever grew. Recent delicate measurements from satellites have shown that the earth is, in fact, very slightly pear-shaped.

Aristotle quoted the circumference of the earth as 400,000 stades, but gave no details of how this figure was found. This is the earliest known estimate. We do know how Eratosthenes (who probably lived from 276 B.C. to 192 B.C.) arrived at his estimate [91]. He knew that at Syene in Egypt the sun was overhead at midday in midsummer: it shone straight down a well. He measured the angle between the zenith and the sun at midday in midsummer in Alexandria, which is practically due north of Syene, and made it one-fiftieth of a complete revolution. (This is about 16' too small. The error might come from using a vertical pole instead of an instrument with a circular scale. See page 28.) He quoted a distance of 5,000 stades between Syene and Alexandria. This makes the circumference 50 × 5,000 = 250,000 stades. Later writers quoted his result as 252,000 stades. They may have wanted the figure for converting stades to degrees. For a 250,000 stade circumference the conversion factor is $694\frac{4}{9}$ stades per degree, and it would be natural to round this figure up to 700. And 700 stades per degree would be equivalent to a circumference of 252,000 stades [92].

How accurate was Eratosthenes's estimate? It is difficult to be absolutely sure, mainly because there were a number of different stades and we do not know for certain which one Eratosthenes used. Pliny says that his stade was one-eighth of a Roman mile [93]. If so, 250,000 stades would be 46,000 km. (Modern measurements make the circumference close to 40,000 km.) In any case, Eratosthenes's figure for the angle between the latitudes of Alexandria and the tropic was 4% too small; the correct value was almost exactly one forty-eighth of a revolution, not one-fiftieth, so his result would be 4% out, even if the distance were dead accurate.

It is, however, possible that the question of Eratosthenes's accuracy is pointless, because the distance from Syene to Alexandria was calculated from astronomical measurements (using an estimate of the size of the earth) and not measured directly. If this is so, then Eratosthenes's reasoning is circular; he has been pulling himself up by his own bootstraps.

The evidence for this suggestion, which is due to Dennis Rawlins [94], is a description of the Nile in Strabo's *Geographica* which Strabo described as "Eratosthenes's opinions":

The Nile is 1,000 stades west of the Arabian gulf and has the shape of a reversed N. It flows 2,700 stades north from Meroë, turns toward the winter sunset, and continues for 3,700 stades, which brings it near to the latitude of the Meroë region. Then, penetrating far into Africa, it turns again and flows 5,300 stades north to the great cataract. Turning a little eastward it covers 1,200 stades to the minor cataract at Syene, and another 5,300 stades to the sea. . . . 700 stades above the confluence of the Astaboras and the Nile is Meroë, a city with the same name as the island. . . . The sea-coast of Egypt stretches 1,300 stades from the Pelusiac mouth to the Canobic mouth.

Not only is the unusual number 5,300 repeated, but 2,700 is half of 5,300 (if rounded off to the nearest hundred) and 700 is a quarter of 2,700. Thus the north–south distances seem to be based on a large unit that can be repeatedly halved. Terrestrial surveying will not produce this result, but astronomical surveying will. A convenient unit would be one-twelfth of the earth's circumference or half, quarter, etc. of this. The north–south distance from Syene to the Mediterranean is close to $\frac{1}{48}$ of the circumference; it is therefore eminently possible that this was the basic unit. Someone converted the distances into stades; to do this he would need an estimate of the earth's circumference. This basic unit must be between 5,250 and 5,350 stades in order to give 5,300 when rounded off. But if it were less than 5,300, half of it would be 2,600, not 2,700. Therefore it must be between 5,300 and 5,350, and so the estimate of the circumference must have been between 254,400 and 256,800 stades.

This estimate of the circumference is between 18 and 19% too high. Rawlins points out that the ancient geographers could have estimated the earth's circumference by finding out how far away a light-house of known height is visible. Because of refraction this method will give a result about 20% too high.

If the map were indeed based on astronomical measurements, that would answer the awkward question of where Eratosthenes's figure of 5,000 stades came from. Although the dimensions of small-holdings along the Nile were precisely known, there is no evidence that the Egyptians computed distances between widely separated places from them, and pacing such distances directly across the sandy desert would have been utterly impracticable.

Rawlins suggests that Eratosthenes took his measurements from the map, under the impression that they were obtained by pacing or some similar method, and rounded off the 5,300 to 5,000 (in which case he must have been working in decimals, not sexagesimals). In that case, of course, all he would find is the original estimate, less 6% because of the rounding, plus 4% because of the inaccurate angle, and so his result would be about 18% too high. And, indeed, it is.

Posidonius (135–50 B.C.) used a somewhat similar method. Rhodes is almost exactly due north of Alexandria and the star Canopus is so far south that when seen from Rhodes it does not rise, but just grazes the horizon. Posidonius measured the maximum altitude to which it rises at Alexandria and found that it was one-forty-eighth of a complete revolution. (The result should have been about one-sixtieth of a revolution.) Cleomedes, after explaining Posidonius's theory in careful detail, said that the distance between Alexandria and Rhodes is about 5,000 stades "or let us assume that this is so" and concluded that the circumference of the earth is 240,000 stades if Rhodes is 5,000 stades from Alexandria "if not, then in proportion to the distance." The distance from Alexandria to Rhodes is actually 570 km, or about 3,650 of the stades we mentioned above. Strabo remarked that some navigators give the length of the sea-passage from Rhodes to Alexandria as 5,000 stades, others as 4,000 [95]. However, Eratosthenes calculated the distance and made it 3,750 stades. Now *if* we used a figure of 3,750 stades in conjunction with the angle of one-forty-eighth of a revolution, we would get a circumference of 180,000 stades [96]. But this would be illogical, because Eratosthenes needed a figure for the circumference to calculate the Alexandria–Rhodes distance. One cannot use the circumference to calculate the distance *and* the distance to calculate the circumference. Strabo may not have realized this because he quoted a figure of 180,000 stades for the earth's circumference as Posidonius's estimate; and this figure has found its way into many later works of reference.

The great classical encyclopedia of geography is Ptolemy's (about A.D. 150). He explained clearly that he used a stade equal to $\frac{2}{15}$ of a Roman mile, and found 500 stades equal to a degree. The figure of 500 stades to a degree corresponds to a circumference of 180,000 stades. This is equal to 192,000 of Eratosthenes's stades. It is an unfortunate coincidence that the actual number, 180,000, that appears is the same as the one Strabo quoted for Posidonius.

Because of this, many modern commentators have blamed Posidonius and Ptolemy for the fact that Sioux, Haidas, and Chickasaws are today called "Indians." Columbus thought that the distance across the ocean from Europe westward to Asia was about 4,500 km, and when he found land at that distance he called the inhabitants Indians. The name stuck, and was applied to all the natives of the New World. However, there were a number of reasons for Columbus's mistake, and the Greek under-estimate of the size of the earth was not the main one. The main reason was an over-estimate of the number of degrees of longitude occupied by the region from the Canary Isles to Japan: Columbus made it 292° (instead of 150°) leaving only 68° instead of 210° for his voyage. When he converted these degrees into miles he used a conversion factor due to Caliph al-Ma'mun, 3° = 170 miles. However, this referred to *Arabic* miles which were in fact some 33% longer than Roman miles, and this made Columbus's estimate fall even shorter of the truth [97].

Does the Earth Move?

Aristotle went to some pains to prove that the earth was at rest. Such a proof has to be philosophical, theological, or psychological; it cannot be astronomical or geometrical. It makes no difference, for example, to the motions of the heavenly bodies relative to the earth whether we regard the earth as rotating eastward or the heavens as rotating westward. (We could even regard them as *both* rotating as long as the difference between their motions is just right.) The first person to regard the heavens as fixed was Heraclides, who may have attended some of Aristotle's lectures. (If so, he was evidently not convinced by them.) According to Simplicius [98]:

Heraclides supposed that the earth is in the center and moves in a circle while the heavens are at rest. He believed that this theory would account for the phenomena.

Aristarchus (about 310–230 B.C.) went even further. He suggested that the earth revolves round the sun. Our evidence comes from Archimedes (287–212 B.C.), that very Archimedes who splashed his way home calling "eureka" after solving the problem of the golden crown. In his book *Psammites* [the sand-reckoner] he cited the universe as an example of something large and in the course of describing it he mentioned Aristarchus's theory, namely, that the sun and the stars are at rest, and the earth revolves in a circular orbit with the sun in the center. He said nothing about whether the planets revolve around the earth or the sun or neither [98a].

Aristarchus

Besides his speculations on the motion of the earth, Aristarchus wrote a more strictly astronomical book, *Peri megethon kai apostematon heliou kai selenes* [On the sizes and distances of the sun and moon]. It is the earliest complete treatise on an astronomical topic that has come down to us from ancient Greece. While astronomy has been making the modest progress we have noted, another science—geometry—has been going ahead by leaps and bounds. The Greek geometers, on the one hand, solved difficult problems, and, on the other, organized geometry as a logical system. The system starts from an explicit list of basic results called axioms that are taken for granted. The idea was that:

(i) these axioms should be so obvious that no one would hesitate to assume that they were true; and
(ii) all geometry should be deducible from them.

Aristarchus's book is arranged in a similar way. He started from six assumptions:

1. The moon gets its light from the sun.
2. The earth is at the center of the moon's sphere.
3. When the moon appears to be halved, the great circle which separates the lit part from the unlit part points toward our eye.

4. When the moon appears to be halved, its [angular] distance from the sun is one-thirtieth of a right angle less than a right angle.
5. The width of the earth's shadow is twice the width of the moon.
6. The apparent size of the moon is one-fifteenth of a sign of the zodiac.

From these, Aristarchus obtained, through a chain of deductions, three results:

I. The distance of the sun is between 18 and 20 times that of the moon.
II. The sun's diameter has this same ratio to the moon's.
III. The diameter of the sun is between $6\frac{1}{3}$ and $7\frac{1}{6}$ times as great as the diameter of the earth.

The mathematics is correct but the basic numbers are badly wrong. The *one-thirtieth* in Assumption 4 should be about *one five-hundredth*, and the *one-fifteenth* in Assumption 6 should be about *one-sixtieth*. These account for the inaccuracy of the final result. Aristarchus also used an assumption that he did not list, namely, that the apparent sizes of the sun and the moon are the same.

Hipparchus

Astronomy took a great leap forward when ingenious geometrical constructions were combined with accurate numerical data. The man who combined them is Hipparchus. Many of his figures for the periods of the moon and planets are the same as ones used by the Babylonians. Although Babylonian astronomy is older than Greek astronomy, the tablets containing the accurate figures are either undated or date from after the conquest of Mesopotamia by the Greeks under Alexander the Great, and it is not clear whether the figures originated with the Greeks or the Babylonians. I will return to this point later (page 128).

The Greeks needed an efficient arithmetic, and this they did take from the Babylonians: they used the decimal/sexagesimal system. However, they did not use cuneiform symbols; they used Greek letters:

$$\alpha = 1, \quad \beta = 2, \quad \gamma = 3, \quad \delta = 4, \ldots,$$
$$\iota = 10, \quad \kappa = 20, \quad \lambda = 30, \ldots, \quad \rho = 100, \quad \text{etc.}$$

Thus what we would write as

$$13° \ 4' \ 32''$$

would appear in a Greek astronomical text as

$$\iota\gamma \ \delta \ \lambda\beta$$

However, the Greeks used a pure decimal system for whole numbers. Thus 141° would appear in a Greek text as

$$\rho\mu\alpha$$

whereas a Babylonian would write it as

$$\text{II} \quad \langle\!\langle \text{I}$$

i.e., as 2,21. I will separate the integral part of a number from its sexagesimal fractions by a semicolon.

The Greeks used simple fractions like $\frac{1}{2}$, $\frac{1}{3}$, $\frac{1}{4}$, etc., as an alternative to sexagesimals when they were more convenient. For example, $\frac{3}{4}$ might sometimes be written as $\frac{1}{2} + \frac{1}{4}$, at other times as 0;45.

Hipparchus's commentary on Aratus and Eudoxus is the only work of his that has survived. It contains some interesting numerical data. For example:

> The head of Ursa Minor lies on the parallel to the equator at the end of Scorpio. When it culminates, so does the third degree of Sagittarius on the ecliptic, and the seventeenth degree of Aquarius rises . . . where the longest day is $14\frac{1}{2}$ hours.

Giving the length of the longest day was a common way of describing latitude, and one well-known place where the longest day was $14\frac{1}{2}$ hours is Rhodes. Scorpio is the eighth sign of the zodiac, so its end has longitude 240°; the reference to the parallel to the equator means that the equatorial coordinate analogous to longitude, i.e., the right ascension, is 240°. By the first degree of a sign Hipparchus meant its beginning (i.e., what we should call 0°, not what we should call 1°). Therefore, the third degree of Sagittarius is 242°, while the seventeenth degree of Aquarius is 316°. In modern language:

The right ascension of the head of Ursa Minor is 240°. For an observer in the latitude of Rhodes, it culminates when 243° on the ecliptic culminates and 316° is rising.

(A star culminates when it is at its highest point as it circles the pole.)

Pliny remarked in his *Natural History* (first century A.D.) that as a result of seeing a "new star" Hipparchus set out to "enumerate the stars for posterity" [99]. This seems to suggest that Hipparchus might have compiled a star catalogue. (The medieval view, derived from Aristotle, was that the heavens were perfect and unchanging. Evidently Hipparchus did not hold this view.) The "new star" that he saw was probably a comet that appeared in 134 B.C. and returned in 124 B.C. It was recorded by the Chinese.

The rest of our information about Hipparchus's work comes from Ptolemy's *Almagest* (about 160 A.D.) [100].

Hipparchus used a "dioptra" to find the apparent diameter of the sun. We learn from Pappus's commentary on the *Almagest* that this consists of a wooden square that can be slid toward or away from a peep hole until it just seems to cover the sun. He calculated the distance of the sun by a method which Ptolemy himself used later. He suspected that the shapes of the constellations did not change but could not prove this for lack of a long series of observations. His own notes contained such details as:

there is a star in line with the Twins' heads and east of them. Its distance from the nearer head is three times the distance between the heads.

These three stars were still in line in Ptolemy's time, 300 years later; this and similar evidence convinced Ptolemy that the pattern of the constellations does not change.

The Length of the Year

Hipparchus measured the length of the year by timing solstices and equinoxes. He first wanted to check that the length was constant—a sensible precaution because, after all, the length of the month varies, and the year might well do the same. And indeed it does, though only by just over $\frac{1}{100}$ of a day. (For example, the time interval between the summer solstices of 1990 and 1991 was 365.2403 days, whereas between the summer solstices of 1991 and 1992 it was 365.2465 days. Modern mathematical astronomers have devised a theoretical length for the year with the small fluctuations, which are mostly caused by a phenomenon known as "nutation," averaged out. The year A.D. 1 was 365.242187 days long, and the year decreases by 0.000006 days per century.)

In Hipparchus's time it was well known that the year was very close to $365\frac{1}{4}$ days. Solstices and equinoxes were reported to the nearest quarter of a day: the report would give the date, and would specify the time as morning, midday, evening, or midnight. Consequently, if one autumn equinox occurred at noon, the next would occur in the evening 365 days later, which will be the same date to anyone using a 365-day calendar. Ptolemy (who did use a 365-day calendar) quoted the following times for autumn equinoxes observed by Hipparchus in the third Callipic cycle (which began in 178 B.C.).

Year 17: sunset
Year 20: morning, 1 day later
Year 21: sixth hour, 1 day later
Year 32: midnight, 3 days later
Year 33: morning, 4 days later
Year 36: evening, 4 days later.

He also quoted three spring equinoxes:

Year 32: morning
Year 43: midnight 2 days later
Year 50: evening 4 days later

We know from modern calculations that these were not accurate [101]. The autumn equinoxes were too late by 17 hours, 12 hours, 12 hours, 7 hours, 7 hours, and 2 hours, respectively. The spring equinoxes were too early by 7 hours, 7 hours, and 5 hours. Hipparchus's timing evidently improved as he went along. But let us look at things from his point of view: Are these data consistent with a year of fixed length?

An observation quoted to the nearest quarter of a day can be out by up to one-eighth of a day. Allowing for this, we can calculate that the autumn equinoxes are consistent with a year of fixed length, but this length would have to be less than $365\frac{17}{76}$ days. This is not consistent with the spring equinoxes, for which the year would have to be greater than $365\frac{17}{72}$ days. The difference is small and, whether or not he made similar calculations, Hipparchus concluded that there was no substantial evidence for any variation.

(The calculation goes as follows. The calendar difference between the quoted dates for the first and last autumn equinoxes is 4 days. If the first equinox was quoted one-eighth of a day late and the last equinox one-eighth of a day early the difference would actually be $4\frac{1}{4}$ days. This is the greatest it can be if the equinoxes are quoted correctly. This difference has taken 19 years to accumulate, so the excess of the year over the calendar year of 365 days could be at most $4\frac{1}{4} \div 19$ days, i.e., $\frac{17}{76}$ days. No other pair of autumn equinoxes give a smaller result. The calculation for the spring equinoxes is similar.)

To find an accurate figure for the length of the year, Hipparchus compared an observation of the summer solstice by Aristarchus in year 50 of the first Callipic cycle with one of his own in year 43 of the third Callipic cycle. These are 145 years apart (a Callipic cycle is 76 years). Ptolemy did not quote the actual times of the solstices, but he did say that the interval between them was half-a-day less than it would have been if the year were exactly $365\frac{1}{4}$ days. Therefore, the year falls short of this figure by 1/290 days, which Hipparchus rounded off to 1/300 days, which is about 5 minutes. (In actual fact, the average year falls short of $365\frac{1}{4}$ days by 11 minutes.)

We do not know how accurate Hipparchus's observation of the solstice was, nor how he made it. Ptolemy described how Hipparchus timed the equinoxes (at least, those that occurred in daylight): he set up a brass ring in a plane parallel to the celestial equator. When the shadow of one-half falls on the other half, the sun is in the plane of the equator and therefore at an equinox.

Periods of the Moon

Before quoting Hipparchus's figures, Ptolemy quoted results known to "the ancients":

$$19,756 \text{ days} = 669 \text{ months}$$
$$= 717 \text{ anomalistic periods}$$
$$= 726 \text{ latitudinal periods}$$
$$= 723\tfrac{4}{45} \text{ sidereal periods.}$$

Indeed, all but the last of these figures were well known; in particular, they occurred in the Babylonian system A. We saw on page 75 that (if we round off 6,585.32 to $6,585\tfrac{1}{3}$)

$$1 \text{ saros} = 6,585\tfrac{1}{3} \text{ days}$$
$$= 223 \text{ months}$$
$$= 239 \text{ anomalistic periods (almost exactly),}$$

and we saw on page 19 that

$$223 \text{ months} = 242 \text{ latitudinal periods (almost exactly).}$$

If we multiply this figure by 3 (to get a whole number of days) we get the figures that Ptolemy quoted, and we see that (except for the sidereal period), Ptolemy was quoting the figures for a triple saros. This period, which the Greeks called an *exeligmos*, is the shortest period that is almost exactly a whole number of days, months, anomalistic periods, and latitudinal periods.

Ptolemy went on to say that Hipparchus improved on these figures. Hipparchus's figures were

$$126,007 \text{ days 1 hour} = 4,267 \text{ months}$$
$$= 4,573 \text{ anomalistic periods}$$
$$= 4,611\tfrac{47}{48} \text{ sidereal periods,}$$
$$5,458 \text{ months} = 5,923 \text{ latitudinal periods.}$$

From these figures we can find the average number of days in a month, etc. Ptolemy said that Hipparchus found

$$1 \text{ month} = 29;31,50,08,20 \text{ days.}$$

In fact, however, if we divide $126007\tfrac{1}{24}$ by 4267 the result is

$$29;31,50,08,09.$$

Hipparchus evidently did not calculate the length of the month from these figures, but simply quoted a previously known result—the same as in the Babylonian system B (page 78). The "5,458 months = 5,923 latitudinal periods" is also the same as in System B. And the "4,267 months = 4,573 anomalistic periods" is simply system B's "251 months = 269 anomalistic periods" with both figures multiplied by 17.

On the question of whether numerical data of this kind originated with the Babylonians or the Greeks, Dennis Rawlins [102] has made an interesting comment. One Babylonian figure for the length of the year is

$$365;14,44,51 \text{ days.}$$

The length of the year is found by counting the days between two summer solstices, or two winter solstices, and dividing by the number of years between them. To get this inaccurate result, the solstice difference must be too big by about 1 day per 240 years. The Greek solstice observations in 432 B.C. by Meton and Euctemon and in 135 B.C. by Hipparchus give *precisely* the Babylonian result, assuming that the solstices were thought to be 108,478 days apart (the correct figure is 108,476.9 days): 108,478 ÷ 297 = 365;14,44,51 to three sexagesimal places. This can scarcely be a coincidence. The correct length of the year was (to three sexagesimal places)

$$365;14,31,55 \text{ days.}$$

Although any reasonably accurate estimate will start with 365;14, only by an amazing coincidence will it end with 44,51.

It is unfortunate that Hipparchus himself used Aristarchus's observation instead of Meton's, with the result that his own estimate (365;14,48) was less accurate than the estimate that (if Rawlins's suggestion is correct) the Babylonians based on his observation.

Table of Chords

For his calculations Hipparchus needed a table in which he could look up the length of the chord *PQ* (see Figure 6.5) if he knew the angle *POQ* and the size of the circle.

We do not know for certain what Hipparchus's table of chords would be like, but G.J. Toomer has suggested that they were calculated for a circle of circumference 21,600 (the number of minutes in 360°) and entries were for $7\frac{1}{2}°$, 15°, $22\frac{1}{2}°$, . . . i.e., for $7\frac{1}{2}°$ intervals. The radius of a circle of this size is 3,438. Toomer's evidence comes from early Indian tables, which were evidently based on early Greek tables [103]. For details of a method by which the table could have been constructed, see Appendix 1.

The Sun's Motion

Hipparchus had to account for the fact that the sun moves irregularly along the ecliptic, that it speeds up and slows down gradually, that it always reaches its greatest speed at the same time of year, and that the time from least speed to average speed is less than the time from average speed to greatest speed. He chose to (or felt he had to) account for these facts using only uniform circular motions. He did this with brilliant

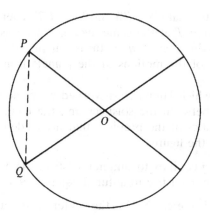

FIGURE 6.5.

simplicity. Instead of combining several motions as Eudoxus did, he used only one, but assumed that the center of the sun's orbit is not the earth but some other point. That is to say, the sun's orbit is *eccentric* in the literal sense of off-center.

In Figure 6.6(a), *C* is the center of the orbit and *T* is the earth (terra). To be able to calculate the sun's position at various times we need to know two basic parameters, the ratio of *TC* to *CA* and the direction of *TC*. We call *TC/CA* the *eccentric-quotient*. *A* is the *apogee*, the point on the orbit furthest from the earth.

If we draw lines through *T* parallel to *CS* and through *S* parallel to *CT*, calling the point where they intersect $\bar{S}$ (see Figure 6.6(b)), we know from

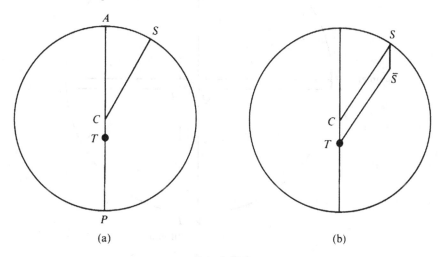

(a)   (b)

FIGURE 6.6

elementary geometry that $CS = T\overline{S}$ and $S\overline{S} = CT$. Therefore $\overline{S}$ revolves at a uniform rate round T at the same average speed as the sun. For this reason $\overline{S}$ is called the *mean sun*. It turns out that it plays an important rôle in the theory of the motions of the planets from Hipparchus right through to Kepler.

The data from which Hipparchus worked were the length of the year and the intervals between the solstices and the equinoxes. It turns out from the mathematics of the problem that he needed only two of these intervals. He used the figures

spring equinox to summer solstice: $94\frac{1}{2}$ days,
summer solstice to autumn equinox: $92\frac{1}{2}$ days.

In Figure 6.7, T is the earth; O is the center of the sun's orbit; and TH, TK, and TL are the directions of the sun at the times of the spring equinox, the summer solstice, and the autumn equinox, respectively. That is, they are the directions from the earth of these equinoctial and solsticial points of the ecliptic. Therefore HTK and KTL are right angles. Knowing how long the sun takes to go from H to K and from K to L, it is not hard to compute the basic parameters. For details of this calculation, see Appendix 2. The results of the calculation, if we use Hipparchus's method as described by Ptolemy and Hipparchus's tables as reconstructed

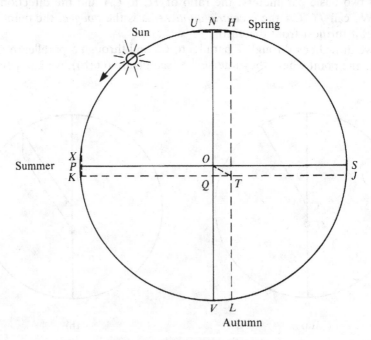

FIGURE 6.7.

by Toomer, are that the eccentric-quotient is 143/3438 and the angle OTH is $65\frac{1}{2}°$: this is the longitude of the apogee. Hipparchus then used these parameters to calculate the lengths of the other two seasons, and ended with the following lengths, in days, for the four seasons:

$$94\tfrac{1}{2}, \ 92\tfrac{1}{2}, \ 88\tfrac{1}{8}, \ 90\tfrac{1}{8},$$

the first two observed, the last two calculated. According to Schiaparelli, the correct figures for 330 B.C. are

$$94.2, \ 92.1, \ 88.6, \text{ and } 90.4;$$

they would not have been very different for Hipparchus's time.

Although the sun does not actually move in an eccentric circle, it does move in a symmetrical curve, and does have an apogee. The longitude of the apogee in Hipparchus's time was $65°68'$, very close to the value that Hipparchus found. Because the sun does not move in an eccentric circle it does not have an eccentric-quotient, but this does not prevent us from calculating what value of the eccentric-quotient would make the eccentric-circle theory most accurate. It turns out to be 0.035, not very close to Hipparchus's result, which is 0.042.

Hipparchus may have been lucky in getting such a good value for the longitude of apogee. The calculated results are quite sensitive to variations in the data, as we can check by repeating the calculation of the basic parameters with the data slightly varied:

94 and $92\frac{1}{2}$ days would give	$61°59'$ and 0.036
$94\frac{1}{2}$ and 92 days would give	$73°49'$ and 0.040
94 and 93 days would give	$54°23'$ and 0.038
$94\frac{1}{2}$ and $92\frac{1}{2}$ days actually gave	$65°30'$ and 0.042
The best values are	$65°58'$ and 0.035.

If we calculate the longitudes of the sun at various times from Hipparchus's results and compare them with longitudes calculated from modern theories, we find that the errors are less than half-a-degree. Most of the error is due to the values of the basic parameters: if we replace these by the best values, the maximum error drops to about 1 minute. The simple theory that the sun moves steadily round an eccentric circle is surprisingly effective.

The Moon's Motion

Hipparchus's theory of the moon's motion is like his theory for the sun, except that the center of the eccentric orbit, D in Figure 6.8, is not fixed but moves in a circle round the earth T. P is the moon, and TD has to rotate at such a rate that the interval between one instant when TDP is a straight line and the next is the anomalistic period of the moon. The same

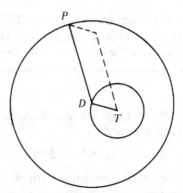

FIGURE 6.8.

effect can be obtained by an epicycle, as shown in Figure 6.9. Here *C* moves in a circle round *T*, while *P* moves in a small circle, the epicycle, round *C*. The dotted lines show why the two systems (which we call the eccentric presentation and the epicyclic presentation) give the same result.

The Greeks regarded the epicycle as fixed to *TC*, like a paper disk stuck to a rod. Then *P* moves once round the epicycle in the anomalistic period of the moon.

In Hipparchus's theory the moon moves clockwise round the epicycle: i.e., *CP* and *TC* rotate in opposite directions. Either direction of rotation round the epicycle would have the same general effect on *TP*—its rate of rotation would speed up and slow down once per anomalistic period.

Some technical terms will be useful. The angle through which the moon has moved round the epicycle, measured from the point on the epicycle

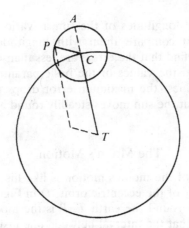

FIGURE 6.9.

furthest from the earth, is called the *epicyclic anomaly*. Thus in Figure 6.9, the epicyclic anomaly of P is the angle ACP. The center of the epicycle is sometimes called the "mean moon": its speed of rotation round T is equal to the average speed of the moon round T, and its longitude is often called the "mean longitude" of the moon. The angle CTP is called the *prosthaphaeresis*: it is the angle that has to be added to or subtracted from the longitude of C to give the longitude of P. (A "prosthesis" is something added, an "aphaeresis" something taken away.) Finally, the circle round which C moves is called the *deferent*.

Accurate observations of the moon run into the following difficulty. Because the moon is not very far from the earth, its direction as seen from one point on earth is slightly different from its direction as seen from another. The difference is called *parallax*. The direction from the center of the earth to the moon is the moon's "true direction." The direction from an observatory X to the moon is the moon's "apparent direction at X."

In Hipparchus's theory T is the center of the earth, not an observatory at Alexandria or Babylon or elsewhere. To avoid trouble with parallax, Hipparchus based his calculations on observations of eclipses. In the middle of an eclipse the moon is dead opposite the sun, whose position is accurately known from the theory of the sun's motion.

As we have seen, Hipparchus had estimates of the average length of a month and of the anomalistic period of the moon. These give the rates of rotation of TC and of CP in the epicycle presentation (and of TD and DP in the eccentric presentation). TC makes one complete rotation in a sidereal period, and CP makes one complete rotation relative to TC in an anomalistic period. To complete the theory we need to find the ratio of TC to CP.

It turns out that Hipparchus needed three eclipses for his calculations. In Figure 6.10, T is the earth, P_1 and P_2 are the positions of the moon at the time of two eclipses, and C_1 and C_2 the positions of the center of the epicycle. In Figure 6.11, the first epicycle has been swung round T to cover the second epicycle. From the time interval between the eclipses Hipparchus could calculate how far round the epicycle the moon traveled between the two eclipses, i.e., the angle P_1CP_2 in Figure 6.11. He could also calculate how far round its orbit the mean moon traveled, i.e., the angle C_1TC_2 in Figure 6.10: call it α. From the times of the eclipses he could calculate the positions of the sun and hence of the moon; from these he could find the angle P_1TP_2: call it β. Then he could find the angle P_1TP_2 in Figure 6.11: it is $\beta - \alpha$.

Using the third eclipse, Hipparchus could similarly calculate angles P_2CP_3 and P_2TP_3 (imagine a third point P_3 on the epicycle in Figure 6.11). From these angles he could calculate CP/TC. For details of this calculation, see Appendix 4. The calculation of TD/DP in the eccentric presentation is closely similar.

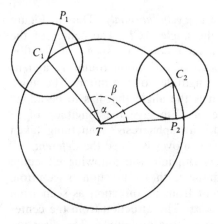

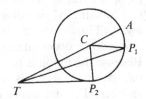

FIGURE 6.10.   FIGURE 6.11.

Unfortunately, the only records we have of data used by Hipparchus
are some that Ptolemy cited to show that Hipparchus's calculations were
wrong, and to try to explain why they were wrong. Hipparchus calculated
TD/DP, using the eccentric presentation, from one trio of eclipses, and
got the value $327\frac{2}{3}/3144$. He calculated CP/TC, using the epicycle presentation, from another trio of eclipses, and got the value $247\frac{1}{2}/3122\frac{1}{2}$.
These are clearly different from each other and different from Ptolemy's
results. (In decimals, they are 0.1042 and 0.0793. Ptolemy got 0.0869 and
0.0872.)

Ptolemy explained this by saying that Hipparchus calculated the time
differences between the eclipses and the corresponding changes in the
sun's longitude incorrectly. The mistakes were small—about 20 minutes
in 178 days and half-a-degree in 175 degrees—but were enough to account for the discrepancy in the final results, which are sensitive to small
changes in the data. Ptolemy was anxious to make it clear that the
difference was not due to the use of the epicycle in one calculation and
the eccentric in the other: the two are equivalent.

To get some idea of the difficulties in calculating the intervals, let us
look at Ptolemy's treatment of the first eclipse. It took place under
Phanostratos archon of Athens, in the month Poseidon. A small part of
the disk of the moon was eclipsed from the north-east when half-an-hour
of the night remained. The moon set still eclipsed.

Ptolemy went on to say that this date was in the year 366 of Nabunasir
on the 26/27th of Thoth "as Hipparchus himself states." The double date
means that the time in question was after midnight on the 26th. From the
date Ptolemy could find the longitude of the sun, and hence the length of
daylight, which was $14\frac{2}{5}$ hours. The "hour" in the phrase "half-an-hour of
the night remained" is one-twelfth of the night. At this date, then, an
"hour" was $1\frac{1}{5}$ regular hours, and the instant when "half of the night

remained" was $5\frac{1}{2} \times 1\frac{1}{5}$ hours, i.e. $6\frac{3}{5}$ hours, after midnight. Ptolemy took this to be the time of the beginning of the eclipse. Since only a small part of the moon was eclipsed, the eclipse could have lasted at most $1\frac{1}{2}$ hours, and so the middle of the eclipse was at $6\frac{3}{5} + \frac{3}{4}$ hours, i.e. $7\frac{7}{20}$ hours, after midnight.

Ptolemy ignored the "at most" and took the duration of the eclipse to be exactly $1\frac{1}{2}$ hours. (Modern tables of ancient eclipses give 102 minutes for this one.) Ptolemy rounded the $7\frac{7}{20}$ hours off to $7\frac{1}{3}$ hours. This is local time in Babylon. To convert it to Alexandria time, Ptolemy subtracted 50 minutes (the correct figure is 58 minutes), making the middle of the eclipse $6\frac{1}{2}$ hours after midnight, Alexandria time. This is 133,250 days $18\frac{1}{2}$ hours after zero-date (see page 141). Finally, Ptolemy reduced this to mean time (see page 142) which changed the $18\frac{1}{2}$ to $18\frac{1}{4}$.

Precession

We now come to the discovery that made Hipparchus famous: the precession of the equinoxes. Ptolemy took his description from Hipparchus's treatise *Peri tes metabaseios ton tropikon kai isemerinon semeion* [On the precession of the solstices and the equinoxes]. Hipparchus found the position of the star Spica by finding the angle between it and the moon at the time of an eclipse. Spica was 6° west of the autumn equinox. But 150 years earlier Timocharis had found Spica 8° west of the equinox. Other stars' longitudes changed at the same slow rate of "not less than 1° in 100 years" (the actual rate is 1° in 72 years), while the latitudes of the stars did not change. Hipparchus concluded that the celestial sphere is rotating, relative to the framework consisting of the equator and the ecliptic, about the poles of the ecliptic.

A Possible Origin for the Constellations

Precession plays an important part in an interesting theory proposed by E.W. Maunder and investigated further by Michael Ovenden: that the Greek constellations were devised, not as pretty pictures, but as a system of coordinates in the sky [104]. Moreover, they were originally devised somewhere between 2000 B.C. and 3000 B.C. by people who lived about 36° north. There are several pieces of evidence for this.

(i) Although constellations come in all shapes and sizes, quite a number are regular enough to lie in a definite direction. In particular, as a star map shows quite clearly, some of the constellations of the zodiac, notably Scorpio, Leo, and Taurus, do not lie along the ecliptic but at an angle to it. Ovenden selected those constellations that were roughly rectangular and appreciably longer than they were wide, and marked out an axis for each. If a constellation were in a plane its axis would be the line perpendicular to the long sides that cuts the rectangle in half (see Figure

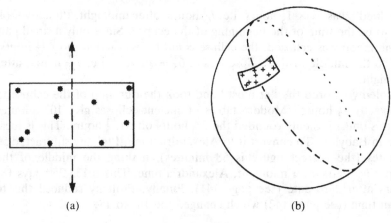

(a)   (b)

FIGURE 6.12.

6.12(a)); because the constellations lie on the celestial sphere, the axes will be the corresponding great circles (see Figure 6.12(b)). Ovenden found that these great circles all passed close to the point on the celestial sphere where the pole was located in 2900 B.C. In particular, Hydra, which is a long thin constellation, lies along the celestial equator of that date. This procedure is highly subjective, so Ovenden asked a colleague to repeat it; the colleague got much the same result. If you want to try it for yourself you will find it quite easy with a celestial globe.

(ii) Ovenden suggested that the information in Aratus's *Phaenomena* dates back to the epoch of the invention of the constellations, and that some of the criticisms that Hipparchus made of it are due to the movement of the pole between that early date and Hipparchus's own time. In particular, at one point Aratus said that Ara "rises over against Arcturus." Hipparchus explained what this means (that the direction of the point on the horizon where Ara rises is opposite to the direction of the point where Arcturus sets) but denied that it was true. And, indeed, it was not true in Hipparchus's time. But about 2200 B.C. it was true. (The conditions that one star rises "over against" another is, of course, that one star is as far from the north celestial pole as the other is from the south celestial pole [105].)

(iii) The risings of the constellations of the zodiac act as a clock. For example, if the sun is in Cancer (as it was at midsummer in classical times) the sun rises when Cancer rises, and so the rising of the previous constellation (Gemini) shows that dawn is near. But we cannot always see Gemini because of clouds, and it is therefore useful to know what other constellations rise (or set) at the same time. Aratus gives, for each of the twelve constellations of the zodiac, a list of simultaneous risings and settings. For example, when Cancer rises:

Corona sets, and so does Piscis Australis as far as its back. Half of Corona is visible, half below the horizon. The waist of backward-facing Hercules is visible but his upper parts are clothed in night. The rising of Cancer brings the unfortunate Ophiuchus down from his knee to his shoulder, and Serpens as far as its neck. Less than half of Boötes is visible: most of him is clothed in night [106].

This means that all these constellations are on the observer's horizon at the same time. This horizon is a great circle on the celestial sphere, and the point on the sphere 90° from it is the observer's zenith. Therefore Aratus, in effect, gives us twelve positions of the observer's zenith. These will lie on a circle whose radius is $90° - \phi$, where ϕ is the observer's latitude, and whose center is the celestial pole. If we know the position of the pole, we know the date of the observations. Ovenden, using statistical methods, found that the latitude was probably between $34\frac{1}{2}°$ north and $37\frac{1}{2}°$ north, and the date between 3400 B.C. and 1800 B.C.

(iv) From the northern hemisphere we cannot see the south celestial pole. In fact, there is a circular region round the pole that we cannot see, and the further north we are the bigger this region will be. To be precise, its radius will equal our latitude. If Hipparchus lived at Rhodes (latitude 31° north) about 150 B.C., the region that he could not see would be a circle of radius 31° whose center is the pole of that date. If the inventors of the constellations lived at about the same latitude, the region that they could not see would be a circle of about the same size but with a different center: the pole at *their* date.

The constellation Eridanus (which represents a river), as described in Ptolemy's catalogue of stars, which is closely based on Hipparchus, flows well into the region which could not have been seen in 2500 B.C. However, there is an interesting constellation map in the Schaubach edition of Eratosthenes's *Katasterismoi* which shows the southern reach of the river quite differently [107]. Instead of flowing into the region of invisibility, it flows along its edge. It is possible, then, that the constellation was originally as shown in this map, but was altered at some time between then and Hipparchus's date. This alteration would account for one of Hipparchus's criticisms of Aratus. Aratus's description of this region of the sky is as follows:

Other small and faint stars circle between Argo's rudder and Cetus and below Lepus. They have no names, because they are not incorporated in any of the constellations [108].

Hipparchus commented that the unincorporated stars are between Argo and Eridanus, not between Argo and Cetus. Aratus must be referring to stars that are not incorporated into the old Eridanus, and these do lie between Argo and Cetus. Hipparchus, however, thought that Aratus was talking about the stars not incorporated into the new Eridanus, and these do lie between Argo and the new Eridanus. These stars, incidentally, are no longer unincorporated; modern astronomers have formed the constellation Columba out of them.

We can add two more pieces of evidence. If Ovenden is correct, the very bright star Canopus (the second brightest in the whole sky, outshone only by Sirius) was not visible to the constellation-makers. And, indeed, Aratus did not mention Canopus. Finally, Aratus did say that Hydra was on the equator [109].

Ptolemy

Ptolemy, who lived and worked in Alexandria, wrote a substantial compendium of early astronomy, the *Almagest*, whose thirteen books occupy nearly 500 pages in modern translation. The word "Almagest" is an Arabic corruption of *Megiste syntaxis* [Greater compendium], the *al* being Arabic for "the." The original title was *Mathematike syntaxeos biblia* ιγ [Mathematical compendium in thirteen volumes]. The dates of Ptolemy's birth and death are not known, but observations he cited as his own range from A.D. 127 to A.D. 141.

The opening chapters set the general scene, perhaps a little dogmatically. Ptolemy did not say that the heavens appear spherical, for instance; he said bluntly that they are spherical. He deduced that the earth is a sphere from three facts. First, the local time recorded for an eclipse is not the same for all observers; the further west the observer, the earlier the recorded time, the time differences being proportional to longitudinal distances between the observers. Because the observers all see the eclipse at the same instant, the difference in local times means that the sun rises later for observers further west. It would not do this if the earth were flat (or concave). Second, observers further north cannot see the more southerly stars. Third, Ptolemy cited the "hull-down" phenomenon (see page 118).

After this logical section comes a contrast. The earth is at the center of the heavens because if it were not, the visible part of the heavens would not be an exact hemisphere. Ptolemy reinforced this reasoning by the Aristotelian argument that the natural movement of a heavy body is toward the center of the heavens. Therefore, if the earth were not already there it would move there. And if the earth is permanently at the center of the universe it must be stationary. He admitted that some people thought it "paradoxical" that the earth does not move, but argued that if it did "animals and other bodies would be left hanging in the air and would quickly fall out of the heavens," and the mere thought of this is ridiculous. He argued similarly against the rotation of the earth: "no clouds, nothing that flies, nothing thrown in the air, could move toward the east; they would all be left behind by the earth and seem to move toward the west." Because the fixed stars have no parallax the earth must be infinitesimal in size compared with the heavens. Ptolemy continued by describing the daily rotation of the celestial sphere and mentioning the complex movements round the ecliptic of the sun, moon, and planets relative to this sphere.

Next, Ptolemy laid down the mathematical background, which consists mostly of what we would today call spherical trigonometry, using, however, not tables of trigonometrical functions but a table of chords (see page 128). For details of this table, see Appendix 3. In the middle of the mathematics comes an interesting piece of astronomy. Ptolemy found the angle ε between the ecliptic and the equator by measuring the angle between the sun and the zenith at a summer solstice and at a winter solstice. The difference between these two zenith angles is 2ε (see page 44). Ptolemy said that he always found 2ε to lie between $47\frac{2}{3}°$ and $47\frac{3}{4}°$, and he remarked that this is about $\frac{11}{83}$ of a complete circle and agrees with a value found earlier by Eratosthenes. (The correct value for 2ε in Ptolemy's time was 47.35°.) Incidentally, this reference in the Almagest is the only evidence we have that Eratosthenes measured the obliquity of the ecliptic: Eratosthenes's work itself has not survived.

The unusual numbers 11 and 83 in this fraction have puzzled many commentators, and various explanations for them have been suggested, nearly all of which depend on supposing that the Greek astronomers started with a measurement in degrees and wanted, for some reason, to convert it to a fraction of a revolution. (Actually, the Greeks used ratios, not fractions, but the two are logically equivalent and fractions are easier for modern readers to follow.) The best explanation is due to D.H. Fowler [111], who started from the two measurements, $47\frac{2}{3}°$ and $47\frac{3}{4}°$, mentioned by Ptolemy. These angles are

$$\frac{143}{1080} \quad \text{and} \quad \frac{191}{1440}$$

of a revolution. ($47\frac{2}{3} \times 3 = 143$; $360 \times 3 = 1080$. $47\frac{3}{4} \times 4 = 191$; $360 \times 4 = 1440$.) We want a convenient fraction that lies between these two fractions, which I call target fractions. (Their average, 229/1728, obviously lies between them but involves uncomfortably large figures.)

We start by noticing that the targets are both between $\frac{1}{8}$ and $\frac{1}{7}$. We continue with a fact known (in its ratio form) to the Greek mathematicians, namely, that if p/q and r/s are fractions then

$$\frac{p + r}{q + s}$$

lies between them. Thus $\frac{2}{15}$ lies between $\frac{1}{8}$ and $\frac{1}{7}$. If it lies between the targets, it is the fraction that we want. If not, it will at least narrow the gap. In fact, it is bigger than both targets. Therefore, the targets lie between $\frac{1}{8}$ and $\frac{2}{15}$: the gap has been narrowed. We continue in this way, trying next $(1 + 2)/(8 + 15)$, i.e., $\frac{2}{23}$. We have the following results:

$\frac{3}{23}$ is too small, so the targets lie between $\frac{3}{23}$ and $\frac{2}{15}$. Next try $\frac{3+2}{23+15}$, i.e., $\frac{5}{38}$.

$\frac{5}{38}$ is too small, so the targets lie between $\frac{5}{38}$ and $\frac{2}{15}$. Next try $\frac{5+2}{38+15}$, i.e., $\frac{7}{53}$.

$\frac{7}{53}$ is too small, so the targets lie between $\frac{7}{53}$ and $\frac{2}{15}$. Next try $\frac{7+2}{53+15}$, i.e., $\frac{9}{68}$.

$\frac{9}{68}$ is too small, so the targets lie between $\frac{9}{68}$ and $\frac{2}{15}$. Next try $\frac{9+2}{68+15}$, i.e., $\frac{11}{83}$.

$\frac{11}{83}$ lies between the targets.

The Sun

The theory of the sun's motion in the *Almagest* is Hipparchus's theory (see pages 128 to 131). Ptolemy started by finding the length of the year. To do this, he compared an observation of the autumn equinox by Hipparchus with one of his own. The result is astonishing—Ptolemy obtained exactly the same result as Hipparchus: $365\frac{1}{4} - \frac{1}{300}$ days. He confirmed his result by comparing two spring equinoxes, and yet again from two summer solstices, the first one a very old one (432 B.C.) attributed to Meton and Euctemon. Ptolemy was unable to improve on Hipparchus, even though he lived some 300 years later.

If we check on the accuracy of Ptolemy's observations we find, to our surprise, that the equinoxes are more than a day out, and the solstice is out by a day and a half. Christian Severin, about A.D. 1600, was the first person to suggest the obvious explanation—that Ptolemy did not observe the equinoxes and the solstices but calculated their times from the times of the three earlier observations, using Hipparchus's value for the length of the year [112]. And if you do just that—start from an autumn equinox at the time cited by Hipparchus, calculate when the autumn equinox 285 years later should occur, taking a year to be $365\frac{1}{4} - \frac{1}{300}$ days, and round the result off to the nearest hour—you will get *exactly* the time and date cited by Ptolemy. The same applies to the other equinox and the solstice.

This suggestion explains another puzzling point. Ptolemy got the same result for the longitude of the sun's apogee as Hipparchus did, namely, $65\frac{1}{2}°$. The correct value in Hipparchus's time was 66°. By Ptolemy's time the longitude had changed to 71°, so Ptolemy should not have found $65\frac{1}{2}°$ again. But if he calculated his equinoxes and solstices from Hipparchus's data, then he would inevitably get Hipparchus's result. (The fact that the longitude of apogee changes was not discovered until about A.D. 1000.) He also obtained the same eccentric-quotient, but using his own tables he found it in the form $2\frac{1}{2}/60$, not the complicated figure reconstructed on page 131.

Why was Ptolemy so eager to reproduce Hipparchus's results that he stretched the truth in order to do so? No one knows. But this desire to confirm existing results could also explain Ptolemy's result for the obliquity of the ecliptic, which agrees with Eratosthenes but not with reality [113].

Ptolemy's work on the sun ends with tables from which the longitude of the sun at any time can be found. He chose midday on the first day of the first Egyptian month (Thoth) in the first year of the Babylonian king Nabu-nasir, a time which was earlier than any observation he used, to act as zero-time—an astronomical "H hour D Day"—the technical name for which is "epoch." Ptolemy made a list of Babylonian kings and the lengths of their reigns, from which we can tell that the year in question is 747 B.C.

Quoting the epoch in terms of a Babylonian year and an Egyptian month may seem peculiar, but there is a good reason for using Egyptian rather than Babylonian months: Egyptian months were all the same length, which makes calculation with them easy. Ptolemy always quoted Babylonian dates in this way. He also quoted many Greek dates, including all the observations he attributed to himself or to Hipparchus, using Egyptian months; Greek months were irregular, and the Greek astronomical calendar would have been hard to use, while the various Greek civil calendars would have been almost impossible [114]. Ptolemy was by no means alone in using Egyptian months; for example, a parapegma from Miletus recorded the summer solstice (actually of 109 B.C.) as both Skirophorion 14 (using an Athenian month) and Payni II [115].

Because of the inaccurate value for the length of the year, Ptolemy's tables lose accuracy by about 6 minutes per year. Because his equinox observations were really based on Hipparchus's, his tables are accurate for Hipparchus's time, not his own. By his time they are 30 hours out, which means that longitudes are about 1° out: if you calculate the longitude for the sun for a date in Ptolemy's lifetime using his tables, the result will be about 1° too small.

Days are not all equally long. The reason is that midday is defined to be the instant at which the center of the sun crosses the meridian, and the time between one crossing and the next varies a little because of the irregular speed of the sun in its orbit. Times of astronomical observations are quoted as so many hours after midday, whereas tables are based on evenly flowing time. We therefore define a *mean solar day* to be the average of the actual days. If we imagine a fictitious "sun" moving evenly round the celestial equator (note carefully—the equator, not the ecliptic), going exactly once round in a tropical year, and passing through the spring equinox point at the same time as the sun, then time as measured by this fictitious "sun" will flow evenly. We call it *mean solar time*. Ptolemy was able to convert times to mean solar times, though he did not use that term: he called them "times accurately calculated."

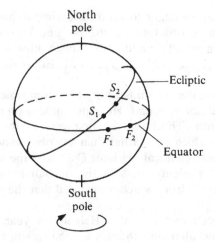

FIGURE 6.13.

Let us suppose that at a certain instant the sun is at S_1 (see Figure 6.13) and the fictitious "sun" at F_1, and that at a second instant the sun is at S_2 and the fictitious "sun" at F_2. The right ascension is, the reader will remember, the celestial coordinate which measures how far round the axis from pole to pole a body has moved. The whole celestial globe is spinning once a day round the axis, and the difference between time as measured by S and time as measured by F is due to the fact that the right ascensions of S and F change by different amounts. Because S and F move in the opposite direction to the rotation of the sphere, we have

time measured by F − time measured by S
= (change in right ascension of S − change in right
ascension of F) ÷ 360,

if angles are measured in degrees and time in days.

We can calculate the right ascension of the sun at the two instants; let us call its increase between the two instants α. Because F moves round the equator at the same speed as the *mean* sun round the ecliptic, its increase in right ascension is the increase in *mean* longitude of the sun: call it $\bar{\lambda}$. Therefore, to convert a time interval to mean solar days we add to it

$$(\alpha - \bar{\lambda})/360 \text{ days.}$$

The greatest value of this correction is just over half an hour.

The correction made to sun-time in order to get mean solar time is sometimes called "equation of time." This comes from the Latin word "equatio" meaning the process of making things equal, and hence coming to mean the adjustment that has to be made to one in order to make it equal the other. The equation of time is the correction that has to be made to a sundial in order to make it agree with a good clock.

The Moon

Ptolemy started by describing Hipparchus's theory of the motion of the moon (see pages 131 to 134) and calculated the radius of the epicycle twice, once from three ancient eclipses and once from three eclipses that he himself observed. He got practically the same answer, namely, 5;13 and 5;14 on a scale in which the radius of the deferent is 60, and concluded that the round number, 5;15 (i.e., $5\frac{1}{4}$), would do.

Next Ptolemy turned his attention to the latitude of the moon (its angular distance from the ecliptic). We have already seen (page 13) that the moon moves in a plane inclined at an angle of about 5° to the plane of the ecliptic. Does the fact that the moon's orbit is not in the ecliptic affect the calculations we have been making? The short answer is *no*. In fact, the greatest possible error in longitude that could be caused by foreshortening is 7'—quite negligible [116].

The points where the moon's deferent (the circle on which the center of the epicycle travels) cuts the ecliptic are the nodes, A and D in Figure 6.14. The line AD rotates slowly in the plane of the ecliptic about its center T (which is the center of the earth) making one revolution in 18.6 years. U is the northernmost point on the deferent. The node A through which the moon passes while moving from south to north is the ascending node. The other is the descending node. The epicycle is in the plane of the deferent, its center is C.

Ptolemy checked the latitudinal period from two partial eclipses just over 615 years apart. In both eclipses the moon was equally far round its epicycle (though on opposite sides) and so equally far from the earth. Moreover, the magnitudes of the eclipses were the same, so the moon was equally far from the node. The moon was eclipsed on the south side, and hence was north of the node, in both cases. Consequently, the moon had made a whole number of revolutions between the two eclipses. From the distance of the moon round the epicycle Ptolemy could calculate the

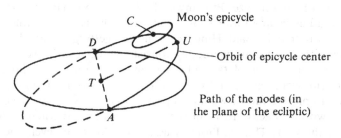

FIGURE 6.14.

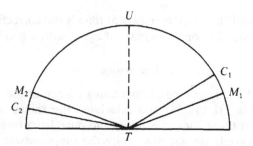

FIGURE 6.15.

angle between the moon and the center of its epicycle as seen from the
earth. It turned out that the moon was 5° behind the epicycle-center at the
first observation and 4°53' ahead of it at the second. Therefore the moon
had covered 9°53' more than the epicycle-center, and so the epicycle-
center had covered 9°53' less than a whole number of revolutions. But
according to Hipparchus's figure for the latitudinal period the epicycle-
center would have covered 8,253 revolutions less 10°2' in the interval
between the eclipses, so Hipparchus's figure is out by 9' in just over 615
years. Ptolemy corrected it by this minute amount.

Next, Ptolemy wanted the value of UTC at his zero-date. He used two
partial eclipses with the moon near the ascending node in the first case
and the same distance from the descending node in the second case, as
shown in Figure 6.15, where M_1 and M_2 are the two positions of the
moon. (C_1 and C_2 are the corresponding positions of the center of the
epicycle.) From the interval between the eclipses and the latitudinal
period Ptolemy found that the center of the epicycle covered 160°4' more
than a whole number of revolutions, from which it is easy to calculate
that UTC_2 = 80°38'. Knowing the date at which UTC equals 80°38' and
the latitudinal period, it is easy to calculate the value of UTC at zero-
date.

As we mentioned earlier (page 134), Ptolemy criticized two of Hipparchus's calculations. It is ironic that this criticism has enabled a modern
commentator to criticize Ptolemy's. R.R. Newton has calculated the
epicycle-radius from all four trios of eclipses—the two that Ptolemy used
and the two that he said Hipparchus used—getting results 5;13, 5;14,
5;14, and 5;16. These are incredibly consistent: the result is very sensitive
to small changes in the data, and if the time of the middle eclipse of the
earliest trio were changed by as little as a quarter of an hour, the
calculated epicycle-radius would drop to 5;4 [117]. Newton suggests that
Ptolemy worked backward from the answer, as he did when finding the
length of the year. That is, Ptolemy assumed that the epicycle-radius was
$5\frac{1}{4}$ and from this calculated the precise data needed, and adjusted the

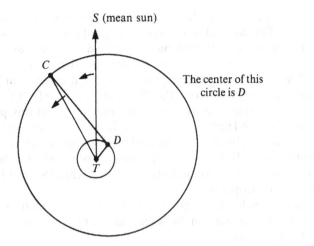

S (mean sun)

C

The center of this
circle is D

D

T

TS rotates at one revolution per year.
TC rotates at one revolution per longitudinal
    period of the moon.
TD rotates so as to keep STD equal to STC.

FIGURE 6.16.

eclipse data accordingly. Why did he choose $5\frac{1}{4}$? Perhaps one calculation was genuine and he adjusted the others to give the same result.

Next, Ptolemy investigated the motion of the moon *between* eclipses. Observations showed that:

(i)   at full moon, Hipparchus's theory is accurate; and
(ii) at half moon the prosthaphaeresis is about 50% greater than the
      theory indicates.

Ptolemy modified Hipparchus's theory to increase the prosthaphaeresis at half moon without affecting it at full moon (or new moon) as follows. The center of the circle carrying C—the *deferent* circle—is not the earth but a point D which revolves around the earth in the opposite direction from the mean moon (see Figure 6.16). When the moon is in conjunction with the mean sun, D is on the line joining them to the earth, and it rotates at such a rate that at the next opposition it is back on the line from the earth to the mean moon. Then halfway between (i.e., at half moon) D is on the opposite side of the earth T from the mean moon C, and so C is nearer to T (by twice the distance DT) than at full moon. This makes the epicycle look larger and so increases the prosthaphaeresis.

This theory combines the eccentric and epicycle constructions: the moon moves round an epicycle whose center moves round a deferent which is not merely an eccentric circle but is a moving eccentric. The rate of rotation of the mean moon C about T is constant, not its rotation about the center D of its deferent. This is a controversial theory because the ancient Greeks had a self-imposed limitation on their theories of celestial motion: all celestial motion should be composed of "regular circular motions." No one knows for certain where this limitation came from—Simplicius said that Sosigenes said that it came from Plato; Geminus said that it came from the Pythagoreans [118]. Nor is a precise definition of "regular" anywhere to be found. Up to this point—as we saw clearly in both Hipparchus's and Eudoxus's theories—regular circular motion has always been the motion of a point which moves round a circle at constant speed. But now Ptolemy uses a motion which is not regular in this sense. Some later astronomers objected to this, the best known of the objectors being Copernicus.

The theory has been set up for full moons and adjusted for half moons, but is still not accurate in between, as an observation by Hipparchus shows. He reported:

longitude of the sun: 37°45'

apparent longitude of the moon: 351°40'

Correcting for parallax, he found

true longitude of the moon: 351°27'30"

From the date of the observation, Ptolemy calculated:

longitude of the center of the epicycle: 352°13'

epicyclic anomaly: 185°30'.

Where exactly should we measure the epicyclic anomaly from? There are two obvious places, A_1 in Figure 6.17(a) and A_2 in Figure 6.17(b). Whichever we choose, M is to the left of C as seen from T and so M has a greater longitude than C, contradicting the figures above. Ptolemy decided not to measure it from either point, but to assume that the moon was at the point M (Figure 6.17(c)) given by its true longitude and to measure the anomaly from the point A, which is 185°30' clockwise round the epicycle from M.

Using Ptolemy's data I find $ACA_1 = 11°51'$ and $A_1CA_2 = 11°59'$ so A is about as far from A_1 as A_2 is but on the other side. Ptolemy had a more ingenious way of locating A. He let N be the point where AC and DT meet, and calculated TN. He found it almost exactly equal to DT. Reversing the procedure he located A as follows: we know how D moves; N is the diametrically opposite point of its orbit; A is the further of the two points in which NC cuts the epicycle.

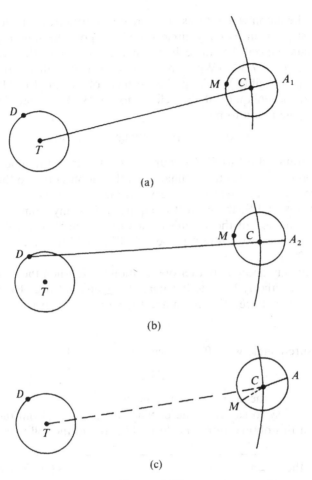

FIGURE 6.17.

The Distance of the Moon

Calculations based on the theory of the moon's motion give its direction relative to the center of the earth; observations give its directions relative to the observer; the difference is parallax (see page 6). Once the parallax of the moon has been found, calculating its distance is an easy piece of trigonometry.

To make his observations, Ptolemy used an apparatus consisting of a vertical rod AB with another rod AC pivoted to it at A. He sighted the moon along AC and measured the distance BC. This gave him the angle CAB, which is the angle between the moon and the vertical.

Before dealing with parallax, however, Ptolemy used this instrument to check on the inclination of the moon's orbit to the ecliptic. To do this, he

observed the moon when it was at its greatest northerly declination and at
the highest point in its daily circle round the pole; he found that under
these circumstances the angle between the moon and the vertical was
always $2\frac{1}{8}°$; i.e., $2°7'30''$. (When the moon is as near as this to the vertical,
parallax is negligible.) Taking the latitude of Alexandria [119] to be
30°58' and the obliquity of the ecliptic to be 23°51', Ptolemy found the
latitude of the moon to be

$$30°58' - 23°51' - 2°7'30'' = 4°59'30'',$$

which he rounded off to 5°. Of course, the latitude of the moon when its
declination is greatest is the inclination of the moon's orbit to the ecliptic.

Ptolemy's measurement is not very satisfactory. The correct latitude of
Alexandria is 31°19', the correct obliquity in Ptolemy's time was 23°41',
and the inclination of the moon's orbit varies from 5° to 5°20'. Thus the
angle which he always found to be $2°7'30''$ should have varied between
2°18' and 2°38'.

Next Ptolemy quoted another observation made when the moon was at
its greatest northerly latitude but near the winter solstice. He calculated
that the angle between the moon and the vertical then was

$$49°48'.$$

The measured angle was 50°55', giving a parallax of

$$1°7'.$$

(This is too big: the correct result is 44'.)

It is now easy to calculate the distance of the moon from the earth at
the time of this observation, and from this, to find the following results:

Radius of the moon's deferent:	48;51,30 earth's radii
Distance from the center of the earth to the center	
of the deferent:	10;8,30 earth's radii
Radius of the epicycle:	5;10 earth's radii

Consequently, the greatest distance of the moon from the earth is 64 radii
and its least distance is $33\frac{1}{2}$ radii. This is a weak point in Ptolemy's theory:
if these distances were correct the diameter of the moon would appear at
times nearly twice as big as at other times, and, as we know, it does not.
Modern values are 65.5 and 55.9 earth's radii, respectively. Ptolemy
would have done better to use a variable epicycle, as Āryabhaṭa was later
to do for the planets (see page 184), increasing its real size instead of its
apparent size. The strong point of the theory is that it predicts the
longitude of the moon accurately, the average error being about 35' once
the all-pervading error due to the incorrect length of the year has been
allowed for [120].

The Distance of the Sun

Having settled the moon's distance, Ptolemy turned to the sun. First, he remarked that the apparent diameter of the moon when it is at its greatest distance from the earth is the same as the apparent diameter of the sun. (This is incorrect. It would mean that the moon's apparent diameter is never less than the sun's. But in annular eclipses a thin ring of sun shows round the rim of the moon, and then clearly the moon's apparent diameter is less than the sun's.) Next Ptolemy compared two partial eclipses, in one of which the moon was one-quarter eclipsed, in the other, one-half. In each case, he calculated the latitude of the moon from its longitude, the longitude of the points where the orbit crosses the ecliptic, and the angle between the orbit and the ecliptic. The difference between the two latitudes, which was 7′50″, corresponds to one-quarter of the diameter of the moon. Therefore the angle α in Figure 6.18 (which corresponds to half the diameter) is 15′40″. Ptolemy also found that the angle β corresponding to the radius of the earth's shadow is 40′40″. From these angles and the fact that the moon's distance is $64\frac{1}{6}$ earth's radii, Ptolemy calculated that the sun's distance is 1,210 earth's radii. (This is far too small: the correct distance is about 24,000 radii.)

This method, which is due to Hipparchus, is so sensitive to small changes in the data, and the data are so hard to measure accurately, that it could lead to almost any result. A little bad luck with the data could even produce a negative parallax.

From his figures Ptolemy constructed tables of parallax for both the sun and the moon.

Eclipses

From the motion of the sun and the moon Ptolemy could calculate the time of full moon and the distance of the moon from the ecliptic then.

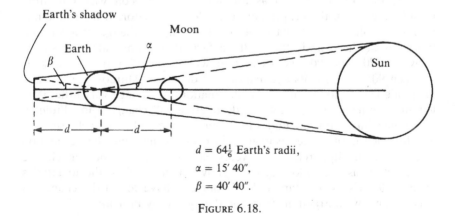

$d = 64\frac{1}{6}$ Earth's radii,

$\alpha = 15'\ 40''$,

$\beta = 40'\ 40''$.

FIGURE 6.18.

Knowing the size of the moon and the earth's shadow, he could calculate
how much, if any, of the earth's shadow is eclipsed then. He went on to
calculate how far from the node (the point where its orbit crosses the
ecliptic) the moon can be at the instant of an eclipse. He found that for
an eclipse of the moon, the mean moon must be within 15°12' of the
node; for an eclipse of the sun it must be between 20°41' one side of the
node and 11°22' the other side. In six months the mean moon travels 184°
relative to the nodes, so that six months after an eclipse there is likely to
be another. In fact, Ptolemy calculated that five months after an eclipse
of the moon there can be another, but only if the five month interval is
one in which the sun is moving round the ecliptic faster than its average
speed, and the moon slower. However, there can never be two eclipses of
the moon seven months apart, even if the moon is moving as fast, and the
sun as slowly, as possible. For eclipses of the sun the results are different:
it is possible for two eclipses of the sun to be visible in the same part of
the world at a seven-month interval.

Ptolemy gave a table from which we can find, for a given full moon,
whether an eclipse occurs then, and, if so, its duration and magnitude. He
also gave tables for eclipses of the sun visible at Alexandria.

The Stars

Ptolemy started his work on the stars by pointing out that they always
keep the same relative positions; he confirmed this by comparing his
observations with Hipparchus's. He went on to describe the precession of
the equinoxes. Hipparchus had given the rate of precession as "at least 1°
per century." Ptolemy compared the longitude of the star in the heart of
Leo (i.e., α Leonis) as calculated from one of his own observations with
its longitude as given by Hipparchus and concluded that the rate of
precession was 1° per century. This is a poor result: the correct value
is 1°23'.

Next, Ptolemy confirmed as follows that precession was a rotation
round the poles of the ecliptic, not the celestial equator. If the rotation is
round the poles of the equator, declinations will not change. If it is round
the poles of the ecliptic they will; in fact, if the longitude of a star is
between 270° and 90° its declination will increase, if its longitude is
between 90° and 270° its declination will decrease. Moreover, if a star's
longitude is near to 90° or 270° its declination will change very little; if its
longitude is near to 0° or 180° its declination will change a lot. Ptolemy
chose nine stars with longitudes in the first range and nine in the second,
and for each quoted the declination in his own time and the declination
in the time of Hipparchus (265 years earlier). The declinations change
as expected, as the following table shows. I have taken the longitudes
from Ptolemy's list of stars and, of course, I have found the change in
declination (by subtraction) from the data given by Ptolemy.

Ptolemy's description	Modern name	Longitude	Change in declination
Shining star in Aquila	α Aquilae	$273\frac{5}{6}°$	2'
*Middle of Pleiades	η Tauri	$33\frac{2}{3}°$	65'
Shining star in Hyades	α Tauri	$42\frac{2}{3}°$	75'
*Capella	α Aurigae	55°	46'
*Orion's western shoulder	γ Orionis	54°	42'
Orion's eastern shoulder	α Orionis	62°	55'
Mouth of Canis major	α Canis majoris	$77\frac{2}{3}°$	15'
First star in Gemini's heads	α Geminorum	$83\frac{1}{3}°$	14'
Second ditto	β Geminorum	$86\frac{2}{3}°$	10'
Heart of Leo	α Leonis	$122\frac{1}{2}°$	−50'
*Spica	α Virginis	$176\frac{2}{3}°$	−66'
*Tip of Ursa Major's tail	η Ursae Majoris	$149\frac{5}{6}°$	−65'
Next star in tail	ζ Ursae Majoris	138°	−90'
Third star in tail	ε Ursae Majoris	$132\frac{1}{6}°$	−81'
*Arcturus	α Boötis	177°	−70'
South claw of Scorpio	α Librae	198°	−94'
North claw of Scorpio	β Librae	$202\frac{1}{6}°$	−84'
Antares	α Scorpii	$222\frac{2}{3}°$	−75'

Ptolemy remarked that from these figures it is possible to confirm the rate of precession, because if it were 1° per century, longitudes would change by $2\frac{2}{3}°$ in 265 years, and from tables he could find for each star what change in declination would follow. He did this for six stars (the ones asterisked), three with increasing and three with decreasing declinations, and said that each time he got a value agreeing with observation. This is nearly true: the figures are:

From the table	65'	46'	42'	−66'	−65'	−70'
Calculated from change in longitude	$63\frac{1}{2}'$	48'	39'	−65'	−64'	−64'

If we calculate the rate of precession for these stars, we get a value close to 1° per century. What of the others? Four are not usable (α Aqu is too near to longitude 270°; α Can, α Gem and β Gem are too near to 90°), but if we use the remaining eight we get an average of 1°28' per century: much better than Ptolemy's result.

Let us look at the values found from each star separately. In the order listed they are:

—	*1°8'	1°32'	*58'	*1°7'	1°46'	—	—	—
1°25'	*1°2'	*1°	1°23'	1°15'	*1°7'	1°30'	1°24'	1°27'

R.R. Newton suggests that Ptolemy wanted to confirm the value 1° and therefore altered the data for the six stars [121]. It is tempting to try to save Ptolemy from this accusation—especially as he has left the data from the other twelve stars for all to see—by suggesting that the data are all genuine and he merely selected the stars which gave values nearest to 1° (i.e., the six smallest values). But there are difficulties with this suggestion. First, it is hard to believe that in a sample of only eighteen measurements, as many as six should be as far out as this. It is remarkable that so many values should be too small, with only one value (from α Orionis) equally far out but too big. And, finally, there is the following statistical argument (though the reader will have to be something of a statistician to appreciate what a strong argument it is).

If you look at a target after target practice you will usually find the good shots grouped closely in the middle, while the bad shots are scattered more widely away from the middle. This is the normal state of affairs for a collection of measurements liable to small random errors: the poor measurements will be scattered more widely than the good ones. Statisticians measure the scatter by a parameter called *variance*. If Ptolemy had selected the six smallest measurements from fourteen genuine measurements, they would have a larger variance than the other eight. In fact, they have a smaller variance.

Ptolemy also quoted the declinations as measured by Timocharis and Aristyllus some 162 years before Aristarchus. To finish the chapter, Ptolemy managed to find observations of three more stars which confirm his wrong value of the rate of precession.

Next comes a list of just over one thousand stars, mostly grouped in constellations, with the longitude, latitude, and apparent brightness of each. The fifteen brightest stars are given a magnitude of 1. Less bright stars have higher magnitudes. The faintest stars listed have magnitude 6, except for a few simply described as "dim." The catalogue lists first the constellations north of the zodiac, then the northern constellations of the zodiac, then (in the next book) the southern constellations of the zodiac, and, finally, the constellations south of the zodiac.

Ptolemy said that he measured the latitudes and longitudes himself. But as long ago as the late-sixteenth century Tycho Brahe suggested that Ptolemy compiled his list by correcting an earlier list (by Hipparchus) for precession, and indeed in Brahe's time the usual way to compile a new list of stars was to modify an existing list. Brahe may well have been right. There are three pieces of evidence that Ptolemy copied at least part of the catalogue from Hipparchus.

(i) Hipparchus worked mostly at Rhodes. Ptolemy worked at Alexandria, which is further south than Rhodes, and so more stars are visible from Alexandria than from Rhodes. Nevertheless, Ptolemy's list contains only stars that were visible from Rhodes in Hipparchus's time.

For instance, α Gruis, slightly brighter than second magnitude, rose 5°
above the horizon at Alexandria in Ptolemy's time and so should have
been visible. Other stars that should have been visible but are not listed
include β Gruis, α Indi and α Phoenicis. (μ Velorum, just over third
magnitude, was visible but was not listed. However, this star was also
visible from Rhodes in Hipparchus's time; I do not know why it was
missed.)

(ii) Surprisingly many longitudes in the list end in an odd $\frac{2}{3}$.

Except for five longitudes which end in $\frac{1}{4}$ and were presumably measured in some exceptional way, all longitudes are either a whole number
of degrees or end in an odd $\frac{1}{6}$, $\frac{1}{3}$, $\frac{1}{2}$, $\frac{2}{3}$, or $\frac{5}{6}$. If they were measured on a
scale graduated in sixths of a degree, the six endings would be equally
common. But not if the scale were graduated in whole degrees and the
odd fractions were estimated by eye. In this case, the measurer may be
reluctant to assign a whole degree to a measurement that does not fall
exactly on the mark, so that anything too small to look like $\frac{1}{3}$ is classed as
$\frac{1}{6}$ even if it is nearer to the visible 0 than to the invisible $\frac{1}{6}$; consequently,
the number of zero endings listed will be below average. Or the measurer
might be more willing to assign a measurement to the visibly marked zero
than to an invisible fraction; if so, the number of zero endings will be
above average. If the scale is graduated in half-degrees, either both 0 and
$\frac{1}{2}$ will be above average or both will be below.

Ptolemy said only that his astrolabon was graduated in subdivisions
of a degree as small as practicable. We do not know how Hipparchus's
astrolabon, if he had one, was graduated.

Latitudes were treated differently; their possible endings included also
$\frac{1}{4}$ and $\frac{3}{4}$. Consequently, latitude endings were not evenly spaced. If the
actual latitudes were evenly distributed and if each was assigned to the
nearest ending, the distribution, in percentages, would be as below. The
actual distribution is shown for contrast.

	0	$\frac{1}{6}$	$\frac{1}{4}$	$\frac{1}{3}$	$\frac{1}{2}$	$\frac{2}{3}$	$\frac{3}{4}$	$\frac{5}{6}$
Even distribution	17	$12\frac{1}{2}$	8	$12\frac{1}{2}$	17	$12\frac{1}{2}$	8	$12\frac{1}{2}$
Actual distribution	23	10	9	10	21	12	4	10

The whole- and half-degree endings are above expectation. This suggests
that the latitudes were measured using a scale graduated in half-degrees
by a measurer more willing to assign a measurement to a visible mark
than to an invisible fraction. This is supported by the fact that it would be
difficult to estimate these particular endings (imagine trying to decide
whether a measurement is nearer to $\frac{2}{3}$ or $\frac{3}{4}$, for example), but much easier
for half-degree graduation—one merely has to estimate whether the
measurement is nearer to one-third, one-half, or two-thirds of the way
along the half-degree interval.

Now let us return to the longitudes. Ptolemy's value for the precession since Hipparchus's time was $2\frac{2}{3}°$. If he had updated a list with an excess of zero endings by adding $2\frac{2}{3}°$ to each longitude he would have had a list with an excess of $\frac{2}{3}$ endings. And, indeed, just over half of his list, namely the northern constellations together with the southern half of the zodiac, does show such an excess:

	0	$\frac{1}{6}$	$\frac{1}{3}$	$\frac{1}{2}$	$\frac{2}{3}$	$\frac{5}{6}$
northern constellations	17	17	19	8	27	$12\frac{1}{2}$
zodiac south	20	14	14	10	30	10

I have omitted the exceptional $\frac{1}{4}$ endings.

If the original distribution were similar to that of the latitudes and $2\frac{2}{3}°$ were added to each, the endings $\frac{1}{4}$ and $\frac{3}{4}$ would become $\frac{11}{12}$ and $\frac{5}{12}$. No such endings appear in the list. There are three obvious things that could have been done to the $\frac{1}{4}$ endings: count them as $\frac{1}{6}$, count them as $\frac{1}{3}$, or share them between the two. If we count them as $\frac{1}{3}$, and the $\frac{3}{4}$ endings as $\frac{2}{3}$, and then increase each ending by $\frac{2}{3}$ we get:

	0	$\frac{1}{6}$	$\frac{1}{3}$	$\frac{1}{2}$	$\frac{2}{3}$	$\frac{5}{6}$
latitudes modified	20	21	17	10	23	10

similar to the distributions above. This suggests strongly that at least these parts of the list were obtained in this way.

The rest of the longitudes are distributed as follows:

	0	$\frac{1}{6}$	$\frac{1}{3}$	$\frac{1}{2}$	$\frac{2}{3}$	$\frac{5}{6}$
southern constellations	25	21	17	11	16	10
zodiac north	26	16	19	10	23	3

(iii) The longitudes in Ptolemy's list are, on average, $1°$ too small. This is just what would happen if Ptolemy updated Hipparchus because the actual precession between his time and Hipparchus's is $3\frac{2}{3}°$, not $2\frac{2}{3}°$.

Argument (ii) is not unassailable; there are other ways of accounting for the preponderance of the odd $\frac{2}{3}°$. For example, Ptolemy could have measured the latitudes and longitudes using a star such as Spica (longitude $176\frac{2}{3}°$) as a reference star (see page 38), setting ring D of the astrolabon to $177°$, the nearest degree to Spica's longitude, and correcting by subtracting $\frac{1}{3}°$ from each of his measured longitudes.

In fact, it is possible to deduce something about the reference stars used. If all the stars in a sizeable constellation used a reference star whose measured longitude is $x°$ too small, the error would be incorporated in the measured longitude of each individual star. The other individual errors, some positive, some negative, would, with luck, on the average, more or less cancel, leaving the average error for the constellation close to the error for the reference star. Conversely, if the average error for the constellation is $x°$, it may well have used a reference star with an error of $x°$. Combining this approach with the supposition that constellations using the same reference stars are likely to be in the same area of the sky, Włodarczyk made the following suggestion [123]:

(1) Reference star Aldebaran: Cancer, Leo, Hydra, Corvus, Procyon, Crater, Cassiopeia, and the northern half of Orion.
(2) Reference star Spica: Virgo, Libra, Scorpio, Corona Borealis, Hercules, Ophiuchus, Serpens, and Sagittarius.
(3) Reference star Regulus: Argo, Centaurus, Lupus, the south part of Corona Australis, and Lyra.
(4) Reference star Fomalhaut: Pisces, Cetus, Eridanus, and Andromeda.
(5) Reference star α Arietis and possibly others: eleven northern constellations, one southern, Aries, Taurus, Gemini, and Aquarius.

Figure 6.19 shows the distribution of these groups on the celestial sphere.

Attempts have been made to controvert argument (iii) by suggesting an alternative explanation for the 1° error, namely, that Ptolemy set his astrolabon using the sun's calculated longitude (see page 38). This longitude was 1° too small. Because the astrolabon is set by pivoting the inner rings about the axis HK (Figure 1.21), the 1° error in longitude would result in the ecliptic ring (ring C) being tilted with respect to the true ecliptic. The tilt would be of the right amount to make this particular longitude 1° too small; other longitudes would be out by different amounts, and latitudes would be affected too. The error in each star's latitude and longitude can be calculated, and this pattern of errors does not show up in the catalogue; consequently, the alternative explanation is not valid [123a].

Ptolemy's treatment of the stars continues with a detailed description of the appearance of the Milky Way and a description of a model of the celestial sphere with the Milky Way and the constellations marked on it, and finishes with definitions of such terms as heliacal rising and culmination—in rather poetic language: Ptolemy's term for heliacal rising is, literally, "early morning east wind"—and some of the relevant spherical geometry.

The Planets

The rest of the *Almagest* is about the planets. First, their order: Ptolemy remarked that all the leading astronomers agreed that the sphere of the fixed stars is outermost, with Saturn, Jupiter, and Mars in that order next,

FIGURE 6.19. Ptolemy's constellations centered on the north celestial pole. A =
Aldebaran, F = Fomalhaut, R = Regulus, S = Spica. Regions 1 to 5 are
explained in the text.

and the moon innermost. They did not agree, though, about Venus and
Mercury. The earlier astronomers thought that they were inside the
sphere of the sun; the later astronomers thought that they were outside.
Ptolemy agreed with the earlier theories, although because there were no
detectable parallaxes, no actual distances could be found.

Ptolemy's theory of the motion of the planets is going to be (with a few
details temporarily omitted) that the planet P moves at a constant speed
round a moving circle, called an *epicycle*, whose center C moves at
constant speed round a fixed circle with center at the center T of the
earth: see Figure 6.20.

If the planet is Venus or Mercury, the point C moves round the earth
in precisely 1 year. In fact, as is clear from tables in the *Almagest*, the
direction TC is always the direction from the earth to the mean sun. P

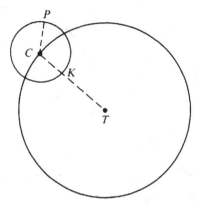

FIGURE 6.20.

goes round the epicycle in one synodic period of the planet; more precisely, if K is the point at which TC crosses the epicycle, the planet moves from K round the epicycle to the new position of K in one synodic period. A comparison with the description on pages 22–23 of how Venus appears in the sky shows that the theory does explain what is seen.

With modern hindsight we can see why the theory works, at least approximately. Today, we consider the motion of Venus or Mercury relative to the earth to be very nearly regular on a moving orbit which is very nearly a circle centered on the sun, which itself moves very nearly regularly on an orbit which is very nearly a circle centered on the earth.

If the planet is Mars, Jupiter, or Saturn, the center C of the epicycle moves round the sun not in a year but in the sidereal period of the planet, and it is the line CP that is always parallel to the line from the earth to the mean sun. Again, we can see why this gives a good approximation to the motion of a planet. Relative to the earth, the planet moves in a large very nearly circular orbit round the sun, while the sun moves in a smaller orbit round the earth: see Figure 6.21(b). If the point C in Figure 6.21(c) is placed so as to make TC equal in length to SP and parallel to it, which will also make TS equal and parallel to CP, then P moves round C, while C moves round T in one sidereal period, as is shown more clearly in Figure 6.21(d).

This theory is not accurate enough to explain the planets' motions in detail. For instance, it would make retrogressions always take the same length of time, whereas, as we saw on page 21, they do not. Ptolemy's technique is to keep the motion of P round C quite regular and to account for all the irregularities by modifying the motion of C. This modification is the "few details temporarily omitted" mentioned above.

The phrase "synodic period" has now changed its meaning. Before any geometrical theory was devised it meant the time interval between the

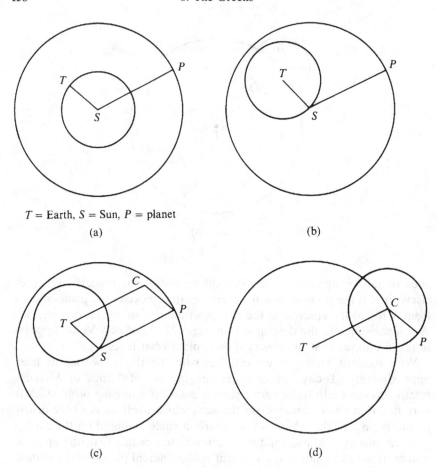

T = Earth, S = Sun, P = planet

(a)   (b)

(c)   (d)

FIGURE 6.21.

beginnings of two successive retrogressions (or their endings, or any
of the other synodic phenomena). This varied slightly, and the early
astronomers found very accurate average values. In the epicycle theory,
the synodic period is the time for the planet to go once round the
epicycle. This is constant, and is taken to be the average time found from
observations. Although Ptolemy did not specifically say so, there is little
doubt that his first modification was to allow C to move on an eccentric
circle, and I shall take this as a starting point in the following description.

Ptolemy needed, of course, the periods of the planets. He quoted them
as follows:

Planet	Synodic period	Time taken in years	Revolutions of the planet
Saturn	57	59 ($+1\frac{3}{4}$ days)	2 ($+1°43'$)
Jupiter	65	71 ($-4\frac{27}{30}$ days)	6 ($-4°50'$)
Mars	37	79 ($+3\frac{13}{60}$ days)	42 ($+3°10'$)
Venus	5	8 ($-2\frac{3}{10}$ days)	8 ($-2°15'$)
Mercury	145	46 ($+1\frac{1}{30}$ days)	46 ($+1°$)

It is interesting to compare these with the Babylonian figures (page 80). Here are the periods, in modern decimals, computed from the two sets of data. While we are about it, we might as well compare them with modern figures (bearing in mind that the periods might have changed a little over the past 2,000 years).

	Synodic period			Sidereal period		
	Babylonian	Ptolemy	Modern	Babylonian	Ptolemy	Modern
Saturn	1.03516	1.03517	1.03517	29.444	29.432	29.639
Jupiter	1.09207	1.09210	1.09211	11.861	11.858	11.860
Mars	2.13534	2.13537	2.13532	1.881	1.881	1.881
Venus	1.6	1.59874	1.59860			
Mercury	0.31724 or 0.31725	0.31726	0.31731			

The modern figures and those for Ptolemy are quoted in tropical years. The Babylonians did not distinguish between the two kinds of year. If their figures are regarded as being in sidereal years they can be converted to tropical years by multiplying by 1.00004.

Mercury

Obviously, the first thing to do is to find the basic parameters of the orbit of the center C of the epicycle, and Ptolemy started with the direction of apogee (the point on the orbit furthest from the earth). We can estimate the distance of C from the earth if we can estimate the size of the epicycle. The sidereal period of Mercury is one year, which means that if C is at a certain position on a certain date, then in a year's time it will be back in the same position. If we select a large number of past observations of Mercury at instants separated by whole numbers of years, we will have a number of observations of Mercury with the epicycle in the same position, and if we have enough of these we can pick out the westernmost and easternmost of them: P_W and P_E in Figure 6.22. The angle PTS

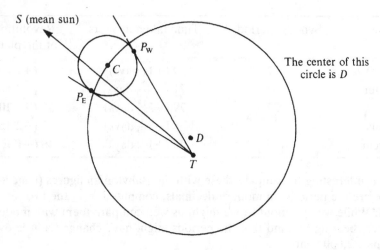

FIGURE 6.22.

between P and the mean sun S is called the *mean elongation* of P, and so when Mercury is at P_W it is at "maximum western mean elongation" for the given position of C. Ptolemy must have had plenty of these records, as he quoted a number of maximum mean elongations. Clearly, the sum of the maximum eastern and western mean elongations is the angle $P_W T P_E$, the angular size of the epicycle as seen from the earth T. This gives a measure of the distance CT: the greater the angle, the smaller the distance.

If D is the center of C's orbit so that A is its apogee (see Figure 6.23) then if C_1 and C_2 are symmetrical about the axis TDA, everything in the diagram is symmetric about this axis, and, in particular, the maximum western mean elongation for C_1 equals the maximum eastern mean elongation for C_2. Ptolemy assumed that the converse was also true and this enabled him, by searching for equal eastern and western maximum elongations, to find the direction of apogee [124]. He found that its longitude was 190°, and that about 400 years earlier it had been 186°, so it had changed at a rate of 1° per 100 years. This is the same as Ptolemy's value for precession, and he concluded that the axis of the orbit of Mercury's epicycle-center is fixed relative to the constellations.

If apogee is at 190°, then 10° must be perigee. But a snag now appears: Ptolemy found that the angle subtended by the epicycle when the longitude of the mean sun was 10° was *less* than when it was 310°. Thus C is nearer to the earth at 310° (and at the symmetrical position 70°) than at 10°, so 10° cannot be perigee. This means that the original theory of C's motion is not valid and must be modified. Ptolemy modified it by introducing a crank mechanism rather like the one he introduced for the moon.

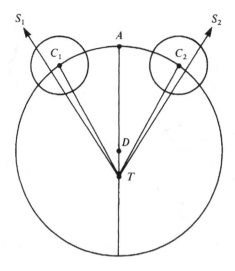

FIGURE 6.23.

In Figure 6.24, T is the earth, D and E are points on the line from T to apogee A, and S is the mean sun. F rotates in a circle around D at the same rate as S around E but in the opposite direction, so that the angle ADF is always equal to AES. The dashed circle has center F, and C is the point where it is cut by ES. As before, Mercury moves round an epicycle with center C; let us call its radius r. The basic parameters of this theory, which have to be calculated from observations, are the distances

$$TE, DE, DF, FC, \text{ and } r,$$

or, rather, their mutual ratios. To find the ratio of, say, TD to r is the same as finding TD in terms of r (e.g., the statement that the ratio of TD to r is $1:4$ is equivalent to the statement that TD is $\frac{1}{4}r$). Because ratios are much less familiar to modern readers than they were to Ptolemy, I use the second of these two ways of expressing the results.

To find the basic parameters Ptolemy started with two observations when the longitude of the mean sun was $190°$ and $10°$. Then, in Figure 6.25, if P_1 and C_1 are the positions of P (Mercury) and C (the center of its epicycle) at the time of the first of these observations and P_2, C_2 at the second, the angles P_1TC_1 and P_2TC_2 are known, and TC_1 and TC_2 can be found in terms of r, giving TD (which equals $\frac{1}{2}(TC_1 - TC_2)$).

Next, Ptolemy used two observations when the line from the earth to the mean sun was nearly perpendicular to the axis TA. Figure 6.26 shows this, S_3 being the position of the mean sun then, C_3 the position of C, and the other letters as above. The angle AEC_3 is a right angle because EC_3 points toward the mean sun. This gives C_3T and ET in terms of r. It

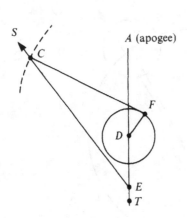

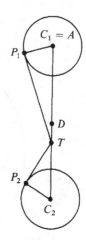

FIGURE 6.24.   FIGURE 6.25.

turned out that ET was very nearly twice TD; Ptolemy took it to be exactly twice, which made ET equal to ED.

In Figure 6.27, F_3 is the position of F at the time of the two observations we are using; DF_3 is also perpendicular to TA. Because DE is small compared with EC_3 (it is about one-twentieth of EC_3) DC_3, EC_3, and C_3T are all approximately equal; we have already found C_3T in terms of r, so now we obtain CF and DF. DF turns out to be close to ED; Ptolemy assumed that they are exactly equal. Ptolemy now had all the proportions he needed. The ratios of the parameters are:

$$TE : DE : DF : FC : r = 3 : 3 : 3 : 60 : 22\tfrac{1}{2}$$

The motion of C is now completely known. To find the motion of P we need to be able to calculate the epicyclic anomaly (which is measured from the point on the epicycle furthest from E) at any time. We can do

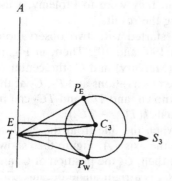

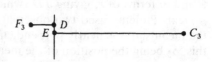

FIGURE 6.26.   FIGURE 6.27.

this if we can find the anomaly at two measured times: the change in anomaly divided by the time interval gives the rate of change of anomaly, and this, together with either one of the measured anomalies, gives the anomaly at any time.

We can find the anomaly at a given time from the longitude. The time gives the position of the epicycle, and a line from T in the direction of the longitude of Mercury will cut the epicycle in two points. If Mercury is at the nearer of these two points, its prosthaphaeresis will be decreasing; if at the further, increasing. Therefore, to find which of the two we want, we need only measure the longitude again a few days later, calculate the prosthaphaeresis, and see whether it has increased or decreased. It is then a straightforward piece of trigonometry to calculate the epicyclic anomaly from the known distances and angles. In this way, Ptolemy found the change in anomaly per day. From this he calculated the anomaly at zero-time, and this completed his work on the longitude of Mercury. The parameters that Ptolemy found gave an average error in longitude of $3°$ and a maximum error of about $7°50'$.

Venus

Ptolemy investigated Venus in the same way as he investigated Mercury, except that, having found that the axis of the orbit of Mercury's epicycle was fixed relative to the constellations, he assumed that this was true also for Venus, and did not use old observations to investigate any possible rotation.

He had no trouble with the perigee and so found no need to modify the original theory by introducing a crank. The theory of Venus's motion is simply as follows (see Figure 6.28). Venus, P, revolves at constant speed round an epicycle of radius r with center C; C moves round a circle center D in such a way that CE rotates at constant speed; E and D are in one straight line with T (the earth). The parameters needed are the ratios

$$TD : DE : DC : r,$$

which Ptolemy found in the same way as for Mercury. In particular, he found that $TD = DE$. This result is important because when he came to investigate the other planets Ptolemy assumed from the start that it was true for them too. His theory would have been more accurate if he had not done so, as modern calculations show, and as Kepler would later discover. It seems an extraordinary coincidence that TD should turn out to be exactly equal to ED, because their relative size is very sensitive to small errors in the data.

How accurate were Ptolemy's results? The average error in longitude is about $1°$, with a maximum error of at least $4\frac{1}{2}°$. If we keep Ptolemy's theory but replace his basic parameters by the ones that make the error smallest (and, in particular, have TD, consequently, not equal to ED),

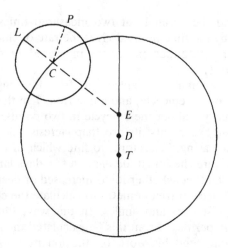

FIGURE 6.28.

the average error is under 10', with a maximum error of about 20' [125]. A point such as E was later called an *equant*: E is an equant for a moving point P if the line EP rotates uniformly round E. Incorporation of the equant in Ptolemy's system increased its accuracy, but violated the principle of regular circular motion (see page 146).

How and why was the equant thought of? Ancient astronomers were well aware that the length of the arc covered by a planet in its retrogression varies according to the position of the arc on the ecliptic. The position where it is longest is diametrically opposite the position where it is shortest.

A simple epicycle would make all retrograde arcs equal. One way to account for the variation is to make the orbit of the mean planet (i.e., the center of the epicycle) eccentric, as Hipparchus did for the sun; if we move the earth to T (Figure 6.29(a)), the retrograde arc as seen from the earth will appear larger near P, the perigee, and smaller near A. (As we will see, the Indians had a different solution. They had an epicycle that varies in size: see page 184.)

So far so good, but the ancient astronomers were also aware that the interval on the ecliptic between the beginnings of two successive retrograde arcs (or the ends of two successive arcs) also varies. This means that the speed of the mean planet round its orbit varies; in fact, it is fastest near P and slowest near A. This irregularity was called the "zodiacal anomaly." In Chapter 10 of Book 6 of the *Almagest*, Ptolemy said that the eccentricity found from the zodiacal anomaly was twice that found from the retrograde arcs. This means that if we make CT the right distance to account for the variation in the retrograde arcs it accounts for only half of the variation in the orbital speed. If we move the point

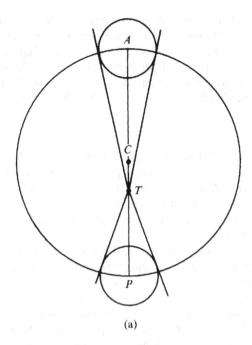

(a)

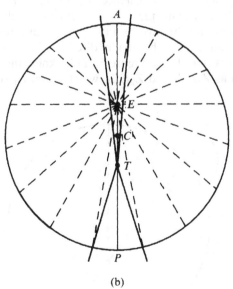

(b)

FIGURE 6.29.

around which the motion is uniform away from T to E (Figure 6.29(b))
the mean planet will move faster near P and slower near A, as shown in
Figure 6.29(b), in which equally spaced radii through E cut off larger
segments of the orbit near P than near A. We might say that the eccentricity found from the retrograde arcs is CT, while the eccentricity
found from the zodiacal anomaly is ET. E is the equant.

Mars

As we have seen, Ptolemy started by assuming that Mars (P in Figure
6.30) moves uniformly round an epicycle with center C, while C moves in
a circle with center D in such a way that if D is the midpoint of ET, then
the line EC rotates uniformly around E. (T is, as usual, the earth.) EC
makes one revolution in a longitudinal period of Mars. CP always points
toward the mean sun.

This means that if Mars is in opposition to the mean sun, C, P, and T
are in a straight line, and so the direction of C is the observed direction of
Mars (see Figure 6.30). Ptolemy started from three such observations. If
the center of the epicycle is at C_1, C_2, and C_3, respectively, then the times
of the observations, together with the longitudinal period, give the angles
C_1EC_2 and C_2EC_3. And the observed positions of Mars give the directions TC_1, TC_2, and TC_3, and hence the angles C_1TC_2 and C_2TC_3 (see
Figure 6.31). By a brilliant piece of mathematics Ptolemy calculated the
eccentric-quotient and the direction of apogee from these data.

This calculation cannot be made directly. The ingenious idea that
Ptolemy used was that if (see Figure 6.32) he knew the angles Z_1TZ_2 and
Z_2TZ_3 (as well as Z_1EZ_2 and Z_2EZ_3, which equal C_1EC_2 and C_2EC_3) he

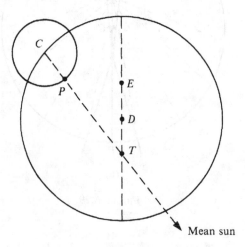

FIGURE 6.30.

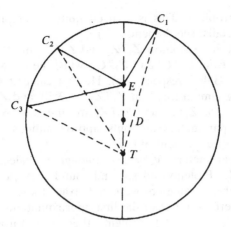

FIGURE 6.31.

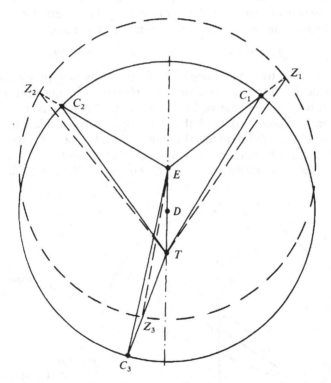

FIGURE 6.32. The dashed circle has center E and the same radius as the deferent circle. Z_1 is the point where the line TC_1 cuts the dashed circle, similarly for Z_2 and Z_3.

could solve the problem. This is in itself quite a complicated piece of geometry. (For details, see Appendix 5.)

Ptolemy did not, in fact, know Z_1TZ_2 and Z_2TZ_3, but they were clearly not very different from C_1TC_2 and C_2TC_3, which he knew from observations (67°50' and 93°44', respectively). He first calculated the basic parameters assuming temporarily that $Z_1TZ_2 = 67°50'$ and $Z_2TZ_3 = 93°44'$. Because his values for Z_1TZ_2 and Z_2TZ_3 are only approximations, these estimates of the parameters are only approximations. He found, for example, an eccentric-quotient of 0;13,7.

Knowing the parameters, it is easy enough to calculate the angles Z_1TZ_2 and Z_2TZ_3. Ptolemy did so, and found them to be 68°55' and 92°21'. Because the parameters were not exact these results were not exact, but they were better than the first approximations (that $Z_1TZ_2 = Z_1CZ_2$ and $Z_2TZ_3 = Z_2CZ_3$). Ptolemy then went through the whole calculation again, using this better estimate of Z_1TZ_2 and Z_2TZ_3 to get a better estimate of the basic parameters. This time the eccentric-quotient turned out to be 0;11,50. From these better-estimated parameters Ptolemy found a better estimate of the angles, namely,

$$Z_1TZ_2 = 68°46' \quad \text{and} \quad Z_2TZ_3 = 92°36'.$$

Yet once more from these angles, which were by now pretty accurate, Ptolemy calculated the basic parameters. The eccentric-quotient this time turned out to be 0;12—that is, $\frac{1}{5}$. Ptolemy judged that by now his results were accurate enough. To check this, he used them to calculate the angles C_1EC_2 and C_2EC_3 and did indeed get the values he started with.

Let Figure 6.33 represent any one of the oppositions. The above calculation gave the ratios between DC, DE, and DT, and the longitude

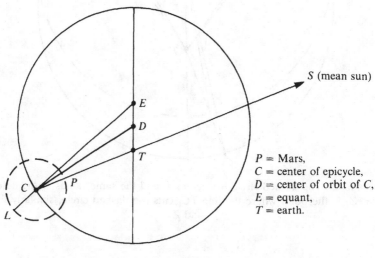

P = Mars,
C = center of epicycle,
D = center of orbit of C,
E = equant,
T = earth.

FIGURE 6.33.

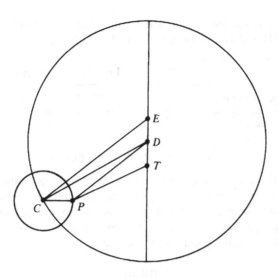

FIGURE 6.34.

of E as seen from T. The longitude of the mean sun—the direction of S as seen from T—gives the angle ETC. From this and the ratio of DC to DT Ptolemy could find the angles of the triangle DCT. This gave him the angle EDC, from which, together with the ratio of DC to DE, he could find the angles of the triangle EDC. Thus Ptolemy found the angle CET, which gives the longitude of C, i.e., the mean longitude of Mars, at the time of the observation. And LCP, the epicyclic anomaly, is $180° - ECD - DCT$. From these Ptolemy easily found the mean longitude and epicyclic anomaly at zero-time.

All that has to be found now is the radius of the epicycle, and any observation when Mars is not in opposition to the mean sun will do for this. In Figure 6.34 let $DC = 60$. Then $ED = DT = 6$. From the time of the observation Ptolemy found the mean longitude and the epicyclic anomaly, which gave him the angles CET and ECP. The longitude of Mars gave the angle PTD. From these, using his tables, Ptolemy found CP. It turns out to be $39\frac{1}{2}$.

Ptolemy's calculations for Jupiter and Saturn use the same method as for Mars.

Accuracy

How accurate are Ptolemy's theories? The errors in longitude are roughly:

	Mars	Jupiter	Saturn
Average error	25′	10′	25′
Greatest error	55′	30′	50′

How accurate is the model itself if we use the best-possible parameters (and allow *TD* and *DE* to be unequal)? Roughly:

	Mars	Jupiter	Saturn
Average error	20'	5'	5'
Greatest error	55'	15'	10' [126]

Calculations from the Theory

Ptolemy continued by computing the greatest and least amounts of longitude lost in an interval of retrograde motion and the time taken. He based his work on a geometrical theorem of Apollonius (third century B.C.). Finally come calculations of the maximum elongations of Venus and Mercury at various points on the ecliptic.

Latitudes

So far, all the calculations have been two dimensional: Ptolemy has ignored deviations from the plane of the ecliptic. The final book deals with these. For Mars, Jupiter, and Saturn, Ptolemy supposed that the plane of the orbit of the center of the epicycle was slightly tilted about the diameter perpendicular to the diameter through its apogee. If we look at the situation for the moment from the usual modern point of view that the sun is fixed, so that the epicycle represents the earth's orbit, we see that the epicycle should be parallel to the plane of the ecliptic. Ptolemy did not know this and had a complicated theory with varying angles of tilt. His theory for Mercury and Venus was even more complicated and Ptolemy remarked (in Toomer's translation) "Let no one, considering the complicated nature of our devices, judge such hypotheses to be over-elaborate . . . we should not judge simplicity in heavenly things from what appears simple on earth." The main reason for the complications is that although the orbit of a planet does lie in a plane, that plane does not go through the earth, as Ptolemy thought, but through the sun.

Note on the Epicycle Theory

In Ptolemy's theory the moon revolves round the center of its epicycle clockwise, while each planet revolves round the center of its epicycle anticlockwise. Why the difference? Epicycles were introduced to account for the speeding up and slowing down of the moon's and the planets' velocities, and either direction of revolution will do this. From the modern point of view, in which epicyclic motion represents motion relative to the earth of a planet revolving round the sun while the sun revolves in the same direction round the earth, the choice is obvious, but this argument would not have made sense to Ptolemy.

However, for a given synodic period, sidereal period, and epicycle radius, a clockwise motion does not slow the motion as much as an anticlockwise one, and, in fact, does not slow it enough to cause retrogression for Mercury, Venus, and Mars. Consequently, if Ptolemy had tried clockwise epicycles for the planets he would have found that they did not work. Mathematical details are in Appendix 6.

Closing Remarks on the *Almagest*

Finally, Ptolemy calculated and tabulated the elongations of the various planets at their heliacal risings. With this the great classical encyclopedia of astronomy comes to an end.

I cannot leave the *Almagest*, however, without saying a little more about R.R Newton's criticisms of it. I have in several places referred to these, but I have by no means covered the whole of Newton's investigation, which occupies a number of research papers and a book, *The Crime of Claudius Ptolemy*. Newton asserts that all the observations that Ptolemy attributes to himself in the Almagest, and many that he attributes to other astronomers, were either invented or modified in order to give the results that he wanted. Newton's thesis, for which the evidence is strong, would largely destroy Ptolemy's reputation as an astronomer, but would still leave us able to appreciate such mathematical *tours de force* as the calculation of the basic parameters for an outer planet from three oppositions.

The Ptolemaic Universe

The planetary theory in the *Almagest* is *angular*. Both inputs and outputs are angles, not distances: it is the angles between heavenly bodies that were assiduously measured and used as data; and it is celestial latitudes and longitudes that theory predicts. (This does not apply to the sun and moon. Ptolemy did calculate their distances.) For any one planet the theory of motion prescribed a certain shape of orbit and, consequently, once the parameters had been calculated, it determined ratios of distances, such as the ratio of a planet's greatest distance from the earth to its least distance. Such ratios are only a byproduct of the theory: in the *Almagest*, Ptolemy made no attempt to check them against observation, or even to speculate whether they agreed with reality.

However, it so happens that although the planets' epicycles were introduced only in order to give correct longitudes, they do correspond to reality: we know today that the orbit of a planet relative to the earth is very nearly a circle around a point (the sun) which itself moves very nearly in a circle round the earth. These two near-circles correspond to Ptolemy's epicycle and deferent (and the fact that they are not quite circles is what prevented Ptolemy from making the earth the center of the deferents). As a result, the shape of the orbit according to Ptolemy's theory is pretty well correct. For the moon the situation is different. Its epicycle was introduced for exactly the same reason—to give correct longitudes—but it does not correspond to reality and gives quite the wrong shape to the orbit.

A second point is that in the *Almagest* the planets were treated separately. Ptolemy just did not deal with such considerations as the ratio of the size of Jupiter's epicycle to Saturn's. The only mention of the solar system as a whole is in Chapter 1 of Book 9, which is concerned with the order of the spheres of the planets. For more details we must turn to a later book of Ptolemy's, *Hypotheseis ton Planomenon* [Planetary Hypotheses] [127]. This book starts by sketching the theory in the *Almagest* with a few small changes in the basic parameters—for example, the radius of Mercury's epicycle was changed from $22\frac{1}{2}$ to $22\frac{1}{4}$. The theory of latitudes was simplified by eliminating the tilting of the planes of the epicycles. The zero date was changed from year 1 of Nabu-nasir to year 1 of Philip of Macedon (324 B.C.). Ptolemy then fitted everything together by making two assumptions. The first was that the moon is nearest to the earth; then Mercury, Venus, the sun, Mars, Jupiter, and Saturn in that order. In reality, this order (except for the fact that the moon is first) was quite uncertain because parallaxes were too small to be measured. If anyone had ever seen Mercury and Venus passing across the face of the sun, that would have shown that they were then closer to us than the sun. The fact that no one saw this did not, however, prove the contrary, because they could be swamped by the sun's glare, making their passage quite invisible. (Even the Chinese, who did see sunspots, never saw Venus or Mercury crossing the face of the sun.)

Ptolemy's second assumption is that no heavenly body trespasses on the territory of another and there is no wasted space between territories. The greatest distance of the moon is 64 earth's radii (Ptolemy rounded $64\frac{1}{6}$ off to the nearest whole number). This, then must be the least distance of Mercury. (In spite of which Ptolemy did not apply parallax corrections to his observations of Mercury.) Ptolemy quoted the ratio of the least to the greatest distances of Mercury as 34 to 88, which makes its greatest distance 166 earth's radii. This will be Venus's least distance, Venus's greatest distance then turns out to be 1,079 earth's radii. The (badly inaccurate) least distance of the sun calculated in the *Almagest* was 1,160. This close agreement, which is of course sheer coincidence, made the system seem correct. So the system continues, until the greatest distance of Saturn turns out to be 19,865 earth's radii.

Ptolemy then quoted observations of the apparent diameters of Venus, Jupiter, Mercury, Saturn, and Mars: they are, respectively, one-tenth, one-twelfth, one-fifteenth, one-eighteenth, and one-twentieth of the apparent diameter of the sun (which is about half a degree). From them the sizes of the planets can be calculated.

Next Ptolemy estimated how far below the horizon the sun must be for a star or planet on the horizon to be just visible: 15° for a first-magnitude star, for example; 7° for Venus at morning setting and evening rising; and 5° at evening setting and morning rising.

In the second volume, Ptolemy considered the mechanism of the motions. Each epicycle is rolled between an inner sphere and an outer sphere. For this we do not need the whole sphere, only a slice through the center thick enough to allow the deviation in latitude. The result is rather like a ball-bearing or roller-bearing. Ptolemy did not, however, connect the spheres of all the planets mechanically, as Aristotle did. In fact, he castigated Aristotle's counter-rotating spheres as senseless. Instead, he compared the planets to a flock of birds, each with its own separate motion.

The dimensions found in the first volume would have provided Ptolemy with an extra argument to show that the sun moves round the earth, not the earth round the sun, had he needed one. If the earth moved, the stars, seen from one position now and from the opposite side of the orbit half a year ago, would show parallax. With an orbit whose diameter is 2,320 earth's radii and stars at a distance of 19,865 earth's radii, this parallax would amount to 7°—easily detectable.

In medieval times the phrase Ptolemaic system referred to this system of nesting orbits; "Ptolemaic" is not synonymous with "geocentric." It is unfortunate that this system was devised, because another method of fitting the separate motions together was crying out to be adopted. Figure 6.35 shows the motions according to the *Almagest*, omitting the details needed to account for small irregularities, namely, the equants and eccentricities. The scales are quite arbitrary. Each planet ties in with the (mean) sun: Mercury and Venus in one way (*TC* always parallel to *TS*), the other planets in another (*CP* always parallel to *TS*). All orbits are presented as epicycles, but of course they could equally well be presented as eccentrics, as in Figure 6.36. Venus and Mercury are now tied in with the sun in the way that the other three were before, and vice versa. This suggests that we might use the epicycle presentation for Venus and Mercury and the eccentric presentation for the others, so that they all tie in with the sun in the same way. Figure 6.37 shows this. In each of them there is an important point (*C* or *D*) in the *same* direction from the earth, namely, the direction of the sun. The distances of these points from the earth are quite unknown. What is more natural than to suppose that all these points are the same point—the sun itself? All we need do is to choose the appropriate scale for each diagram. Then each planet will circle the sun while the sun circles the earth—a coherent system.

What prevented Ptolemy from thinking of this? The answer is probably "crystalline spheres," though, of course, it is possible that he simply failed to see it. We know (page 117) that Aristotle believed that the spheres in Eudoxus's system were real, and the *Hypotheseis* makes clear

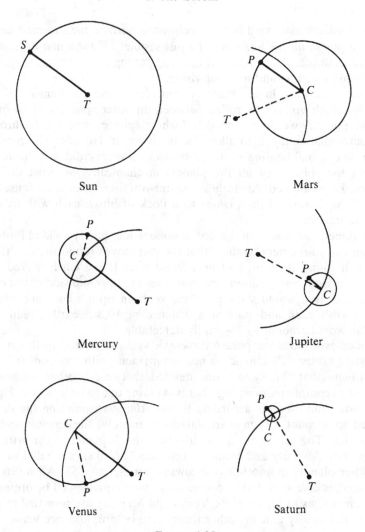

FIGURE 6.35.

that Ptolemy too believed that the spheres carrying the heavenly bodies were real. In fact, European astronomers believed this up to the time of Tycho Brahe (A.D. 1546–1601), who investigated comets and showed that they would have to penetrate any such spheres.

The heliocentric system that I have been describing is shown, not to scale, in Figure 6.38(a). The spheres carrying Mercury, Venus, and even Mars would have to intersect the sphere carrying the sun, which rules the system out. (Actually, for Ptolemy, the spheres in Figure 6.38(a) should be doubled: each body has two spheres, whose radii are its greatest and

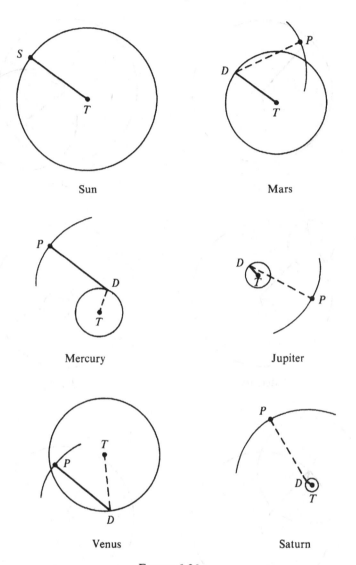

Sun   Mars

Mercury   Jupiter

Venus   Saturn

FIGURE 6.36.

least distances from the earth.) If we consider the earth, not the sun, to move we get Figure 6.38(b), which poses no problems. But belief in the movement of the vast solid earth on which we stand takes sophisticated imagination.

Incidentally, Tycho did suggest a system like that of Figure 6.38(a), but not until well after Copernicus (A.D. 1473–1543) had suggested a system like that of Figure 6.38(b).

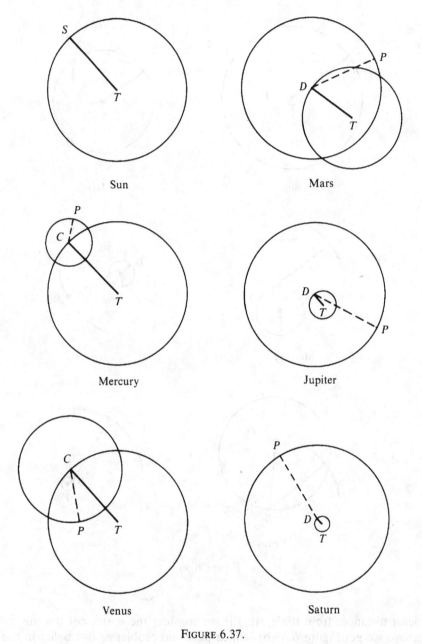

FIGURE 6.37.

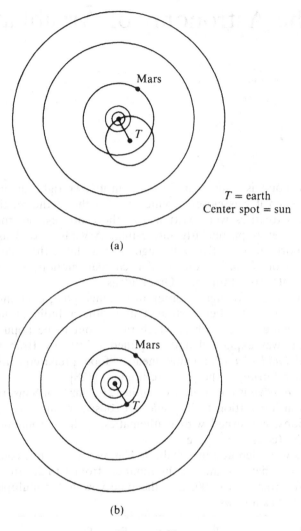

T = earth
Center spot = sun

(a)

(b)

FIGURE 6.38.

Before we leave Ptolemy, we should notice that besides the two astronomical works just described and a third, *Phaseis aplanon asteron*, dealing with heliacal risings and settings and the corresponding weather predictions, Ptolemy wrote an astrological treatise (the *Tetrabiblos*). He also wrote a geographical treatise that is almost as prominent in classical geography as the *Almagest* in classical astronomy.

CHAPTER 7

The Astronomy of Āryabhaṭa

Indian astronomy is a substantial subject in its own right, arising in early antiquity, when it was closely bound up with the Vedic religion. Many astronomical treatises were written over the centuries, but the work of one astronomer is particularly interesting: Āryabhaṭa of Kusumapura, who was born in A.D. 476. Although he was later than Ptolemy, his planetary theories, based on regular circular motion, seem to have developed out of pre-Ptolemaic Greek ideas.

Āryabhaṭa had two slightly different systems. One, expounded in the *Āryabhaṭīya*, was still being studied in southern India as late as the nineteenth century. The other, which turned out to be popular in northern India, was explained and simplified a little by Brahmagupta in the *Khaṇḍakhādyaka* (the name means "food prepared with candy," presumably referring to the pleasure it gives) [128].

The *Āryabhaṭīya* is concisely written in 121 stanzas of Sanskrit verse. It starts with an invocation to Brahmā and consists of an introduction and three sections, concerned with mathematics, with (astronomical) times, and with the (celestial) sphere.

Numbers were denoted by syllables. The first twenty-five consonants of the Sanskrit alphabet stand for the numbers from 1 to 25, the next eight for the tens from 30 to 100, and the nine vowels for multiplication by powers of 100 as follows:

k	kh	g	gh	ṅ	c	ch	j	jh	ñ
1	2	3	4	5	6	7	8	9	10

ṭ	th	ḍ	dh	ṇ	t	th	d	dh	n
11	12	13	14	15	16	17	18	19	20

p	ph	b	bh	m
21	22	23	24	25

y	r	l	v	ś	ṣ	s	h
30	40	50	60	70	80	90	100

a	i	u	ṛ	ḷ	e	o	ai	au
$\times 1$	$\times 100$	$\times 10,000$	$\times 1,000,000$	$\times 10^8$	$\times 10^{10}$	$\times 10^{12}$	$\times 10^{14}$	$\times 10^{16}$

thus gṛ means 3,000,000 and khuyughṛ means 4,320,000.

The number of revolutions of the various celestial bodies in a *yuga* are:

Sun	4,320,000	Mercury	17,993,020
Moon	57,753,336	Jupiter	364,224
rahu	232,226	Venus	7,022,388
Mars	2,296,824	Saturn	146,564

Thus a *yuga* is 4,320,000 years. It is a period at whose beginning and ending these bodies were presumed to have celestial longitude zero. In the same period the earth made 1,582,237,500 rotations (which makes a year equal to 366.2589 sidereal days). *Rahu* is the moon's ascending node, personified as a celestial demon.

The size of the universe and the various orbits are as follows. A *yojana* is a unit of length, 8,000 times the height of a man. The circumference of the sky in *yojanas* is ten times the number of arc-minutes covered by the moon in a *yuga* (i.e., $10 \times 21,600 \times 57,753,336$) and the length of the orbit of a planet is the circumference of the sky divided by the number of revolutions of the planet in a *yuga*. This makes the orbit of the moon 216,000 *yojanas* and means that all planets are moving at the same speed. It also means that the sizes of the orbits are proportional to the periods of the planets. This gives a picture of the universe as a whole, but one based on an incorrect assumption (quite different from the equally incorrect assumption in Ptolemy's *Hypotheseis ton planomenon*). The sizes of the orbits in terms of the sun's are:

Moon	0.075	Venus	0.62	Jupiter	11.9	The sky	4,400,000
Mercury	0.24	Mars	1.88	Saturn	29.5		

The diameters of the earth, the sun, and the moon are 1050, 4410, and 315 *yojanas*. Therefore the radius of the moon's orbit is 65.5 earth's radii. If a man's height is about $1\frac{1}{2}$ meters, the radius of the sun's orbit will be about 5,500,000 km, one-thirtieth of the correct value. The earth's radius will be about 12,600 km, quite accurate.

The angle between the ecliptic and the equator is 24°. The greatest deviations from the ecliptic and the longitude of the ascending node of each planet is listed. The trigonometrical table used in the *Āryabhaṭīya* is a table of sines, presented by listing the differences between successive entries:

225, 224, 222, 219, 215, 210, 205, 199, 191, 103, 174, 164,
154, 143, 131, 119, 106, 93, 79, 65, 51, 37, 22, 7.

By successive addition we can reconstruct Āryabhaṭa's table of sines ($225' = 3\frac{3°}{4}$).

Angle	0	$3\frac{3}{4}°$	$7\frac{1}{2}°$	$11\frac{1}{4}°$	15°	$18\frac{3}{4}°$	$22\frac{1}{2}°$	$26\frac{1}{4}°$	30°
Sine	0	225	449	671	890	1,105	1,315	1,520	1,719

Angle	$33\frac{3}{4}°$	$37\frac{1}{2}°$	$41\frac{1}{4}°$	45°	$48\frac{3}{4}°$	$52\frac{1}{2}°$	$56\frac{1}{4}°$	60°
Sine	1,910	2,093	2,267	2,431	2,585	2,728	2,859	2,978

Angle	$63\frac{3}{4}°$	$67\frac{1}{2}°$	$71\frac{1}{4}°$	75°	$78\frac{3}{4}°$	$82\frac{1}{2}°$	$86\frac{1}{4}°$	90°
Sine	3,084	3,117	3,256	3,321	3,372	3,409	3,431	3,438

The sine of an angle is the perpendicular distance from one end of an arc subtending that angle at the center of a circle of circumference 21,600 to the diameter through the other end of the arc: see Figure 7.1. Because 21,600 is the number of minutes in a complete revolution, the sine of a small angle is approximately the size of the angle in minutes. This table is the one that suggested that early Greek tables of chords were for a circle of this size and at intervals of $7\frac{1}{2}°$ (the sine of an angle is half the chord of twice the angle): see page 128.

Āryabhaṭa's theory of motion for the heavenly bodies is as follows.

The Sun

The *kakṣyāmaṇḍala* or *kakṣyāvṛtta* is a circle with center at the center T of the earth and lying in the plane of the ecliptic. A point $\bar{S}$ moves round this circle at uniform speed in such a way that the line $T\bar{S}$ points toward the sun exactly once a year. We can think of $\bar{S}$ as the "mean sun." It carries an epicycle. The sun S lies on this epicycle in such a way that the

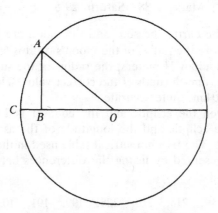

The circumference of the circle is 21,600.
The sine of the angle AOB is AB.

FIGURE 7.1.

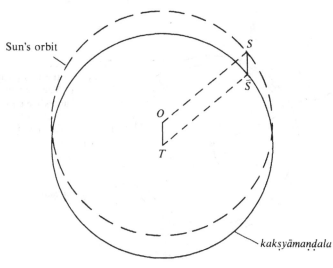

Sun's orbit

kakṣyāmaṇḍala

T: earth.
S̄: mean sun.
S: sun.
O: center of the sun's orbit.

FIGURE 7.2.

line $S\bar{S}$ is always in the same direction in space. This is, of course, the same as the theory of Hipparchus (except for proportions: the circumference of the epicycle is $13\frac{1}{2}$ units, a unit being $\frac{1}{360}$ of the circumference of the *kakṣyāmaṇḍala*; Hipparchus' figure is 15 units). See Figure 6.6(b) and Figure 7.2.

The Moon

The theory for the moon is the same as for the sun, but with a relatively larger epicycle ($31\frac{1}{2}$ units).

The Planets

Let us start with Venus. A point V_s, called the *śīghrocca* of Venus, moves at uniform speed round the earth in such a way that at every superior mean conjunction (i.e., whenever the mean sun is directly between the earth and Venus) the line TV_s points towards Venus, and between two successive superior mean conjunctions V_s moves once-and-a-bit round the earth: see Figure 7.3. It is the *śīghrocca* of Venus (not Venus itself) whose revolutions in a *yuga* were listed above.

A point $\bar{P}$ moves round the *kakṣyāmaṇḍala* of Venus in such a way that $T\bar{P}$ always points to the mean sun. Because the planet appears sometimes

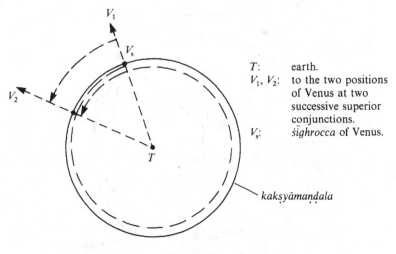

T: earth.
V_1, V_2:   to the two positions
              of Venus at two
              successive superior
              conjunctions.
V_s: śīghrocca of Venus.

FIGURE 7.3.

on one side of the sun and sometimes on the other it will appear sometimes on one side of $\bar{P}$ and sometimes on the other, while $\bar{P}$ moves
uniformly. Consequently, $\bar{P}$ is called the mean [*madhya*] planet [*graha*].
The planet itself is called the *sphuṭagraha: sphuṭa* literally means "clear"
or "evident" but in this context is usually translated as "true."

The point $\bar{P}$ carries two epicycles called *manda* (slow) and *śīghra* (fast),
used for making adjustments to the longitude of $\bar{P}$ in order to obtain the
longitude of the planet. This inevitably reminds us of the Greek theory
that each planet has two anomalies (a zodiacal anomaly and a solar
anomaly) [129].

On the *śīghra* epicycle is a point P_s placed so that $\bar{P}P_s$ is parallel to the
direction from T to the *śīghrocca*. The angle $\bar{P}TP_s$ is the *śīghra* adjustment to the longitude of $\bar{P}$ used in obtaining the longitude of the planet.
(Figure 7.4(b)).

On the *manda* epicycle, which is quite small, is a point P_m placed so
that the line $\bar{P}P_m$ is in a fixed direction in space, the *mandocca*. (The
mandoccas of the various planets are listed.) The angle $\bar{P}TP_m$ is the
manda adjustment. (Figure 7.4(a)).

Because $\bar{P}P_m$ is in a fixed direction, P_m moves round a fixed eccentric
circle. Because $\bar{P}P_s$ points to the *śīghrocca*, the *śīghra* epicycle on its own
is equivalent to the simplified motion shown in Figure 6.20. It allows
for the fact that the planet's orbit is actually centered on the sun, but
does not allow for any irregularity; this is taken care of by the *manda*
epicycle. Thus the two epicycles between them amount to something like
an eccentric-and-epicycle theory which may well be the one used by the
Greeks before they introduced the equant. In fact, if the *śīghra* epicycle

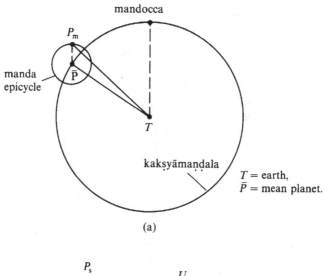

(a)

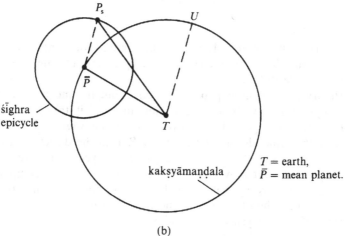

(b)

FIGURE 7.4.

were carried by P_m instead of by $\bar{P}$ the motion would be precisely
that: see Figure 7.5. The size of the *śīghra* epicycle compared to the
kakṣyāmaṇḍala is equivalent to the size of Venus's epicycle compared to
the circle round which its center moves in the Greek theory, and to the
size of Venus's orbit compared to the earth's orbit in real life. The size of
the *manda* epicycle compared to the *kakṣyāmaṇḍala* is equivalent to the
Greek eccentric-quotient. In contrast to the Greek theory, Āryabhaṭa's
epicycles are not fixed: they expand and contract rhythmically as time
passes.

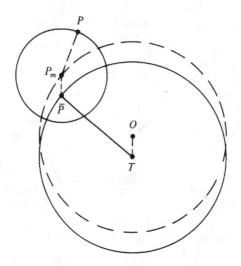

FIGURE 7.5.

The theory of motion for Mercury is the same as for Venus. Of course, the period of the *śīghrocca*, the sizes of the epicycles, and the direction of the *mandocca* are different.

The theory for the outer planets is similar but the mean planet $\bar{P}$ no longer keeps in line with the mean sun; instead, it travels round the *kakṣyāmaṇḍala* in a period of its own. And the *śīghrocca* is a point P_s on the *śīghra* epicycle moving so that $\bar{P}P_s$ is in the direction from the earth to the mean sun. The size of the *manda* epicycle still represents the eccentric-quotient, but now the size of the *śīghra* epicycle compared to the *kakṣyāmaṇḍala* is equivalent to the orbit of the sun compared with that of the planet, the other way round from before.

The sizes of the epicycles are as follows:

	manda	*śīghra*
Moon	$31\frac{1}{2}$	
Sun	$13\frac{1}{2}$	
Mercury	from $22\frac{1}{2}$ to $31\frac{1}{2}$	from $130\frac{1}{2}$ to $139\frac{1}{2}$
Venus	from 9 to 18	from $256\frac{1}{2}$ to $265\frac{1}{2}$
Mars	from 63 to 81	from $229\frac{1}{2}$ to $238\frac{1}{2}$
Jupiter	from $31\frac{1}{2}$ to 36	from $67\frac{1}{2}$ to 72
Saturn	from $40\frac{1}{2}$ to $58\frac{1}{2}$	from $40\frac{1}{2}$ to 46

The longitude of a planet is found by applying the *manda* and *śīghra* adjustments as follows. From the known periods, the longitude of the mean planet $\bar{P}$ at any given time can be found. Then, since the longitude of the *mandocca* A is known, so is the angle $AT\bar{P}$, where T is the earth: see Figure 7.4(a). The Sanskrit term for this angle is *mandakendra*. The *manda* adjustment is the angle $\bar{P}TP_m$. From the known periods we can find the angle $UT\bar{P}$, where U is the *śīghrocca*: see Figure 7.4(b). The Sanskrit term for this angle is *śīghrakendra*. The *śīghra* adjustment is the angle $\bar{P}TP_s$. The *manda* adjustment depends only on the *mandakendra*, not the *śīghrakendra*; the *śīghra* adjustment depends only on the *śīghrakendra*, not the *mandakendra*.

On the face of it, there are three ways in which the two adjustments can be combined. We might (i) simply calculate the two adjustments for the given position of $\bar{P}$, add them, and apply the result. Or (ii) calculate the *manda* adjustment, apply it, calculate the *śīghra* adjustment for the new position, and apply it. Or, equally, we could (iii) calculate first the *śīghra* adjustment and apply it, and then calculate the *manda* adjustment for the new position and apply it. Āryabhaṭa adopts none of these methods, but one considerably more complicated.

For Mars, Jupiter, and Saturn the method described in the *Āryabhaṭīya* is as follows. Let P_1, P_2, and P_3 be points on the *kakṣyāmaṇḍala* placed so that the angle $P_1T\bar{P}$ is half the *manda* adjustment for $\bar{P}$, the angle P_2TP_1 is half the *śīghra* adjustment for P_2, the angle $\bar{P}TP_3$ is the *manda* adjustment for P_2, and the angle P_3TP is the *śīghra* adjustment for P_3. The longitude of P is the longitude of the planet. In other words, we adjust the mean planet by half the *manda* adjustment followed by half the *śīghra* adjustment, which gives an intermediate position P_2. We then start again from the mean planet, adjusting it not by its own *manda* adjustment but by P_2's, followed by the *śīghra* adjustment.

The stanza in the *Āryabhaṭīya* which describes this, shown with a transliteration in Figure 7.6, is remarkably compact. A literal translation is

For the *mandocca* [and] *śīghrocca* half [is] taken negatively [or] positively for the planet and the *manda*. The *sphuṭamadhya* should be known from the *mandocca* and the *sphuṭa* from the *śīghrocca* [130].

The *sphuṭamadhya* is the position called "intermediate" above. The phrase "negatively or positively" refers to the fact that the adjustments some-

मन्दोच्चाच्छीघ्रोच्चादर्धमृणं धनं ग्रहेषु मन्देषु
मन्दोच्चात् स्फुटमध्याः' शीघ्रोच्चाच्च स्फुटा ज्ञेयाः

mandoccācchighroccādardhamṛṇaṁ dhanaṁ graheṣu mandeṣu
mandoccāt sphuṭamadhyāḥ śighroccācca sphuṭa jñeyāḥ

FIGURE 7.6.

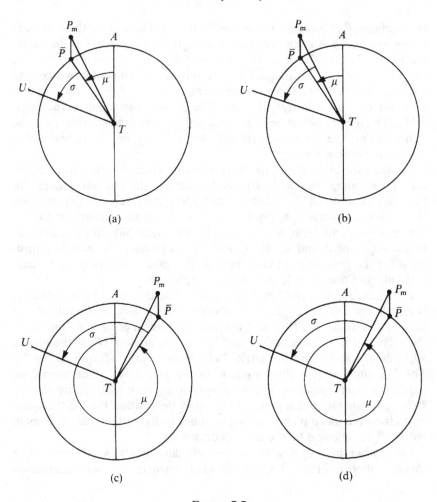

FIGURE 7.7.

times increase and sometimes decrease the longitude. In fact, as is clear from Figure 7.7, the *manda* adjustment decreases the *mandakendra* (and therefore the longitude) and increases the *śīghrakendra* if the *mandakendra* is between 0° and 180°, and acts in the opposite direction if the *mandakendra* is between 180° and 360°. Similarly, the *śīghra* adjustment increases the *mandakendra* and decreases the *śīghrakendra* if the *śīghrakendra* is between 0° and 180° and acts oppositely if not.

For the inner planets Venus and Mercury the procedure is a little simpler: it omits the first (*manda*) half-adjustment.

In Āryabhaṭa's other system, which uses the same procedure for all planets, the first two half-adjustments are made the other way round:

śīghra first, then *manda*. In a commentary on the *Khaṇḍakhādyaka* written in the eleventh century, Varuna gives the following example [131].

The longitude of the mean planet	325°01'10"	
The longitude of the apogee	127°	
The longitude of the *śīghrocca*	168°01'55"	
The *mandakendra* of the mean planet	198°01'10"	i.e., 325°01'10" − 127°
The *śīghrakendra* of the mean planet	203°00'45"	i.e., 168°01'55" − 325°01'10"
The corresponding *śīghra* adjustment	−32°07'37"	from tables
The first adjusted *mandakendra*	181°57'22"	i.e., 198°01'10" − ½(32°07'37")
The corresponding *manda* adjustment	22'50"	from tables
The second adjusted *mandakendra*	182°08'47"	i.e., 181°57'22" + ½(22°50')
The corresponding *manda* adjustment	25'	from tables
The third adjusted *mandakendra*	198°26'10"	i.e., 198°01'10" + 25'
The corresponding *śīghra* adjustment	−31°47'35"	from tables
The *mandakendra* of the planet	166°38'25"	i.e., 198°26'10" − 31°47'35"
The longitude of the planet	293°38'35"	i.e., 166°38'25" + 127°

The reasoning behind these procedures is not obvious and one modern commentator, D.A. Somayaji, has remarked that all interpreters of Indian astronomy consider them peculiar and irrational, and quotes *Bhāskara* as saying:

It is really curious. Tradition has it, so it must be respected. Agreement with observation is the only proof [132].

Otto Neugebauer tried to explain the mysterious use of the halved adjustments by suggesting that applying the *manda* adjustment at opposite ends of the arc corresponding to the *śīghra* adjustment produces errors of opposite sign. This, if true, would suggest applying it in the middle, i.e., using half the *śīghra* adjustment, but unfortunately it is not true [133]. The distance of the planet from earth is the distance TP_s multiplied by TP_m and divided by $T\bar{P}$ (see Figure 7.4(a,b)).

Further Topics

Besides the astronomical theory just described, the *Āryabhaṭīya* contains some spherical trigonometry, a procedure for calculating the magnitude and duration of an eclipse, given the moon's latitude, and a brief description of a model celestial sphere rotated by means of a hydraulic drive. The mathematics section covers various topics in pure mathematics not directly related to astronomy, and has an exceptionally accurate estimate of the circumference of a circle of radius 20,000. The circumference is 62,832, equivalent to estimating π to be 3.1416.

In contrast to all this, the *Āryabhaṭīya* has some relics of primitive cosmology: mount Meru, one *yojana* high, is at the center of the earth and gods living there see the celestial sphere as moving from left to right. In one "day of Brahmā" (1,008 *yugas*) the size of the earth increases by

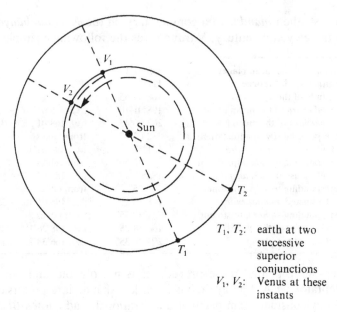

T_1, T_2: earth at two
successive
superior
conjunctions

V_1, V_2: Venus at these
instants

FIGURE 7.8.

one *yojana*; in an equally long "night of Brahmā" this increase is lost. But
there is also the following remarkable passage:

Just as a man in a boat sees the trees on the bank move in the opposite direction,
so an observer on the equator sees the stationary stars as moving precisely toward
the west.

Not only did Āryabhaṭa believe that the earth rotates, but there are
glimmerings in his system (and other similar Indian systems) of a possible
underlying theory in which the earth (and the planets) orbits the sun,
rather than the sun orbiting the earth. The evidence is that the basic
planetary periods are relative to the sun. For the outer planets this is not
significant: both earth and sun are inside their orbits and so the time
taken to go round the earth and the time to go round the sun are the
same. The significant evidence comes from the inner planets: the period
of the *śīghrocca* is the time taken by the planet to orbit the sun. We see
this as follows. Figure 7.8 shows two successive superior conjunctions
relative to the sun. Between these two instants the planet revolved round
the sun one revolution plus the angle covered by the earth in that time,
and this is precisely the angle covered by the *śīghrocca*: see Figure 7.3.

Unwritten Astronomy

It is worth noticing that Indian mathematical astronomy was not confined
to written sources such as the *Āryabhaṭīya*. In 1850 an English officer

found in Pondicherry a calendar-maker who was able to calculate the time of an eclipse of the moon, using procedures that he had learnt by heart, representing numbers by sea-shells laid out in a pattern on the ground [134].

We might wonder how the Indians were able to calculate the times of eclipses with any accuracy in view of their rather simple theory of the moon's motion. The reason is that a simple theory works fairly well for full moons and new moons, as we can see by referring to Ptolemy's theory (pages 143 to 146) in which the complications are introduced in order to give accurate longitudes between new and full moon.

CHAPTER 8

Arabic Astronomy

Greek astronomy came to an end in Europe with the collapse of Hellenistic civilization not long after Ptolemy's time. But it was carried on by the Arabs, or rather by the people of the middle-eastern Arabic-speaking Islamic civilization that arose as a result of Mohammed's conquest in the seventh century A.D. This civilization comprised not only Arabs but Persians and Turks, as well as Moors, Kurds, and others. Writers on architecture use the word "Saracenic" to describe this culture, but historians of science seem to prefer "Arabic." Arabic astronomy is a substantial and specialized subject; here I can deal only with some of the more interesting highlights.

Serious Arabic astronomy started through contact with India. At any rate, the earliest surviving Arabic work that is anywhere near complete is the *Zij al sind-hind* of al-Khwārizmi (born shortly before A.D. 800). *Zij* is an Arabic word often used for astronomical tables and *sind-hind* is an arabicized version of the Sanskrit *siddhānta*. The *zij* used Indian parameters [135].

Two consequences of the contact with India were important not just for astronomy but for mathematics in general. One was the introduction of the Hindu–Arabic numerals

١ ٢ ٣ ٤ ٥ ٦ ٧ ٨ ٩

(still to be seen, for example, on automobile number plates in Cairo), which evolved into our familiar 1, 2, 3, 4, 5, 6, 7, 8, 9, 0. Even sexagesimal calculation is much easier with these digits than with either Babylonian wedges or Greek letters, and Europe eventually changed to pure decimal calculations, as the Chinese had been doing all along, though those astronomers who quote angles in degrees, minutes, and seconds instead of decimals of a degree have not yet completed the change. The other important consequence was the replacement of tables of chords by tables of sines.

190

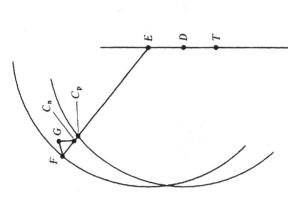

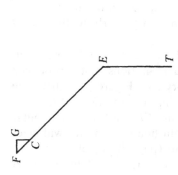

Naṣir al-Dīn's theory of the motion of a planet.

T: the earth.

E: Ptolemy's equant.

EF is of constant length (equal to the radius of Ptolemy's deferent) and rotates around E at a constant speed (in the longitudinal period of the planet).

$FG = GC = \frac{1}{4}TE$.

FG rotates around F twice as fast as EF around E but in the opposite direction, starting with $TEFG$ in a straight line.

GC is parallel to TE. This ensures that the angle EFG is equal to the angle GCF, which ensures that C is on the line EF.

C is the center of the epicycle. It moves uniformly round G, which moves uniformly round F, which moves uniformly round E.

The motion of the planet round C is the same as in Ptolemy's theory.

(a)

Naṣir al-Dīn and Ptolemy compared.

D: center of Ptolemy's deferent.

C_p: center of Ptolemy's epicycle.

C_n: center of al-Dīn's epicycle.

The line from E to the center of the epicycle is the same in both theories. C_p is the point where this line crosses the deferent.

(b)

FIGURE 8.1.

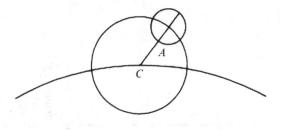

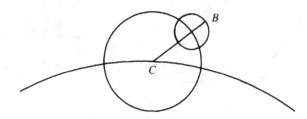

FIGURE 8.2.

The Arabs translated Ptolemy's *Almagest* and other works. Indeed, the very name *Almagest* is a version of the Arabic name *al-majasti* for the work which Ptolemy called the *Syntaxis* [136]. More important, most of the *Hypotheseis ton Planomenon* would have been irretrievably lost, were it not for the Arabic translation [137]. The Arabic theorists did not follow Ptolemy slavishly but disagreed on a number of points. For example, Ibn al-Haytham (A.D. 965–1040) wrote a treatise *Al-shukūk 'ata Batlamyūs* [Doubts about Ptolemy] objecting to motions that could not be produced by a combination of regular circular motions, particularly to the use of the equant [138].

Three centuries later, Naṣir al-Dīn (A.D. 1201–1294), also known as al-Tūsi, devised combinations of regular circular motions that reproduced closely the motion given by Ptolemy's theory (see Figure 8.1(a,b)). The last part of his linkage is particularly interesting: the rotations of *CG* and *CF* (see Figure 8.1(a)) make *C* move along *EF*, and so two regular circular motions combine to produce a straight-line motion. We will meet this again, as it was to be used by Copernicus (page 209) [139].

The theory was modified slightly by his pupil Qutb al-Dīn al-Shirāzi (A.D. 1236–1311) and by Ibn al-Shātir (A.D. 1304–1376), who also devised a greatly improved theory for the moon [140].

Ptolemy's theory for the moon (page 148) gave an obviously incorrect ratio between the greatest and least distances of the moon from the earth.

FIGURE 8.3. Bust of Ulugh Beg. (Photograph by Ernest W. Piini.)

He had found it necessary to increase the apparent size of the epicycle at half moon, which he did by means of a crank that drew the epicycle nearer to the earth then. Al-Shātir instead increased the size of the epicycle: he had the moon moving round a small extra epicycle whose center moves round the epicycle just mentioned at such a speed that at half moon it is at A (see Figure 8.2) and so is effectively on an epicycle of radius CA, and at new moon and full moon it is at B, effectively on an epicycle of radius CB. This device was also used by Copernicus.

Arabic astronomers were also keen observers and measurers. Caliph al-Ma'mun had an observatory built in Baghdad in A.D. 829, Naṣir al-Dīn established an observatory in Maragha in A.D. 1259, and Arabic astronomy culminated in the work of Ulugh Beg (A.D. 1394–1449) (Figure 8.3), whose enormous observatory at Samarkand was described on page

32. He produced tables based on an extremely accurate estimate of sin 1°, and he compiled a catalogue of stars with 1,018 entries.

Arabic astronomers were able to calculate parameters more accurately than the Greeks, partly because of the longer time-spans available, and partly because of their larger instruments. For example, al-Battānī (active at Raqqa between A.D. 878 and 918) found the obliquity of the ecliptic to be 23°35′ [141]; Ulugh Beg found it to be 23°31′17″ [142]. Here, reduced to decimals for ease of comparison, is a summary.

	Approximate date		Error
Eratosthenes	200 B.C.	23.86°	0.12°
Ptolemy	A.D. 140	23.86°	0.17°
Al-Battānī	A.D. 900	23.58°	0.01°
Guo Shoujing	A.D. 1280	23.56°	0.03°
Ulugh Beg	A.D. 1400	23.52°	0.00°
Copernicus	A.D. 1500	23.47°	−0.04°
Brahe	A.D. 1600	23.525°	0.025°

Al-Battānī also estimated the length of the year by comparing the time of the autumn equinox in A.D. 880 with the one recorded by Ptolemy in A.D. 139. His result is $2\frac{1}{2}$ minutes too small (compared with modern estimates of the average length of the year at that date) [143]. Most of the error is due to Ptolemy, who put his equinox a day too late; had he put it on the right day al-Battānī's estimate would have been only half a minute too small. Later, Ulugh Beg estimated the year to be 365 days, 5 hours, 49 minutes, 15 seconds; this is 25 seconds too large [144]. Here is a summary.

			Error
Hipparchus	150 B.C.	365.24667	0.00433
Al-Battānī	A.D. 900	365.24056	−0.00273
Al-Zarqali	A.D. 1270	365.24225	0.00028
Guo Shoujing	A.D. 1280	365.2425	0.00023
Ulugh Beg	A.D. 1400	365.24253	0.00027
Copernicus	A.D. 1500	365.24256	0.00030
Brahe	A.D. 1600	365.24219	−0.00001

Improved parameters made improved astronomical tables possible, especially the Great Hakemite tables (about A.D. 1000) by Ibn Yūnus, the Toledan tables, slightly later, of al-Zarqali, and the widely used Alphonsine tables, introduced in A.D. 1272 and so called because they were devised for King Alfonso X of Spain.

Arabic astronomy played a substantial role within Arabic culture, for example, in accurately determining the direction of Mecca from localities in the wide-ranging Islamic domain; but its main importance for the general history of astronomy lies in:

(i)   the preservation of Ptolemy's work;
(ii) the improved theory of the moon; and
(iii) accurate determination of parameters.

CHAPTER 9

The Mayas [145]

Far away from any of the civilizations we have mentioned, the Mayas flourished in central America between about 300 and 900 A.D. They too studied the sky. We do not know for sure what they thought of the shape of the earth, but the later Aztecs thought that the crust of the earth was the back of a huge alligator, and the Mayas may well have believed the same.

The Maya calendar used a cycle of twenty names:

| Imix | Ik | Akbal | Kan | Chicchan | Cimi | Manik | Lamat | Muluc | Oc |
| Chuen | Eb | Ben | Ix | Men | | Cib | Caban | Eznab | Cauac | Ahau |

Each day was given a number and a name, starting with 1 Imix. The next day is 2 Ik, then 3 Akbal and so on. After 13 the numbers go back to 1, so that the fourteenth day is 1 Ix, and they continue like this until the twentieth day is 7 Ahau. Then the names start again, so that the next day is 8 Imix, and so on. In fact, the 13 numbers and the 20 names fit together in the same way as the 10-character and the 12-character sequences in the Chinese sexagesimal cycle. The Mayan cycle has, of course, a length of 260. All this is, however, only a start.

The Mayas recognized a 365-day unit that we might call the Maya year. (It has the same length as the Egyptian year and, like the Egyptian year, gets out of step with the seasons by one day every four years.) It consists of 18 *uinals*, each of 20 days, plus 5 extra days. The names of the *uinals* are:

| Pop | Uo | Zip | Zotz | Tzec | Xul | Yaxkin | Mol | Chen |
| Yax | Zac | Ceh | Mac | Kankin | Muan | Pax | | Kayab | Cumhu |

The days of each *uinal* are numbered, and this is straightforward numbering, like the modern month-and-day-of-the-month dates: the day after 1 Pop is 2 Pop, and after 20 Pop comes 1 Uo.

FIGURE 9.1. Maya glyph at Dos Pilas, Guatemala. The top part of the glyph is the symbol for Venus; the bottom part is the symbol for Seibal, a nearby town. The glyph commemorates the victory of Dos Pilas over Seibal in a battle thought to have been deliberately timed at an auspicious juncture in the cycle of Venus. (Photograph E.C. Krupp, Griffith Observatory.)

The Mayas specified a date by giving all four components: 1 Imix 4 Pop, for instance. (What would the next day be?)* Archaeologists call this a "calendar-round" date. Calendar-round dates form a cycle of 18,980 days. (18,980 is the least common multiple of 365 and 260.)

A second system of dating, called "long-count," simply numbered the days, starting from a zero date whose calendar-round is 4 Ahau 8 Cumhu and which Mayanists have pin-pointed, by correlating long-count dates with known events, as 3114 B.C., August 13th (Gregorian). The Maya used the *uinal* of 20 days, the *tun* of 360 days, the *katun* of 20 *tuns* and the *baktun* of 20 *katuns*. Therefore

$$9 \text{ baktuns } 14 \text{ katuns } 19 \text{ tuns } 5 \text{ uinals } 0 \text{ days}$$

represents a date $9 \times 20 \times 20 \times 360 + 14 \times 20 \times 360 + 19 \times 360 + 5 \times 20$ days, i.e., 1,404,100 days, after zero date. It turns out to be in A.D. 730. The intermediate units could be omitted and the above date could appear simply as

$$9, 14, 19, 5, 0.$$

*See page 258.

The Maya represented numbers by dots and bars, a bar counting as 5, and used a sea-shell to denote zero, so this number would be written (or carved) as

though it would be arranged vertically, not horizontally.

The Moon

The Mayas often included data for the moon in their records. For example, on a stela at Palenque they inscribed

3846 tuns   108 days   month 5 day 19.

The second part indicates that this is the nineteenth day of the fifth month. The numbers of the months are never greater than 6, so evidently the Mayas numbered them from 1 to 6 and then started again. If they had had occasion to record the date ten days later it would have read

3846 tuns   118 days   month 5 day 29

and the next day would be

3846 tuns   119 days   month 5 day 30

unless a new moon appeared, in which case it would be

3846 tuns   119 days   month 6 day 1.

Some of these moon-data are calculated, not observed. For example, the temple of the foliated cross in Palenque records

765 tuns   80 days   month 5 day 10.

This is a date in 2359 B.C., and is far too old for the moon data to be a record of an observation.

This sets an interesting problem. What figure for the length of a month did the Mayas use? The difference between the two dates is 3,081 *tuns* 28 days which is 1,109,188 days. (A nice easy calculation. Much easier than finding, for instance, the number of days between September 27, 55 B.C., and March 13, A.D. 1975.) The moon is 9 days younger on the second date; this means that 1,109,179 days is a whole number of six-month periods. A six-month period is about 177.18 days. If we divide 1,109,179 by 177.18 we get 6,260.05, so the whole number of six-month periods is 6,260; and if the Mayas' figure for the average length of a month is reasonably accurate they must have found the same result. Therefore their figure must have been such that 1,109,179 days equals 37,560 months, to the nearest day. The Mayas might, of course, have been using the relation

$$1,109,179 \text{ days} = 37,560 \text{ months},$$

but this is not very likely: it would be an amazing coincidence if the difference between the first pair of dates that we happened to look at were one of the figures of the basic relation. And the numbers are uncomfortably large. So we look for two smaller numbers that give approximately the same result. There is a well-established mathematical technique for this (expand the quotient as a continued fraction and examine the successive convergents). It yields the relation

$$2,392 \text{ days} = 81 \text{ months}$$

as a good approximation. If the Maya used this they would find that 37,560 months is $1109179\frac{1}{27}$ days. This is correct to the nearest day. There are other pairs of dates at Palenque on which this ratio can be tested, and it works every time. Therefore we can be reasonably confident that we have found the estimate of the length of the month used at Palenque.

Venus

The Maya were also interested in Venus. The synodic period of Venus (the time for it to go through a complete cycle of appearances) is very nearly 584 days. The Maya split this period into four parts: 236, 90, 250,

Mayan Almanac for Venus.

Cib	Cimi	Cib	Kan	Ahau	Oc	Ahau	Lamat	Kan	Ix	Kan	Eb	Lamat	Etz'nab	Lamat	Cib	Eb	Ik	Eb	Ahau
3	2	5	13	2	1	4	12	1	13	3	11	13	12	2	10	12	11	1	9
11	10	13	8	10	9	12	7	9	8	11	6	8	7	10	5	7	6	9	4
6	5	8	3	5	4	7	2	4	3	6	1	3	2	5	13	2	1	4	12
1	13	3	11	13	12	2	10	12	11	1	9	11	10	13	8	10	9	12	7
9	8	11	6	8	7	10	5	7	6	9	4	6	5	8	3	5	4	7	2
4	3	6	1	3	2	5	13	2	1	4	12	1	13	3	11	13	12	2	10
12	11	1	9	11	10	13	8	10	9	12	7	9	8	11	6	8	7	10	5
7	6	9	4	6	5	8	3	5	4	7	2	4	3	6	1	3	2	5	13
2	1	4	12	1	13	3	11	13	12	2	10	12	11	1	9	11	10	13	8
10	9	12	7	9	8	11	6	8	7	10	5	7	6	9	4	6	5	8	3
5	4	7	2	4	3	6	1	3	2	5	13	2	1	4	12	1	13	3	11
13	12	2	10	12	11	1	9	11	10	13	8	10	9	12	7	9	8	11	6
8	7	10	5	7	6	9	4	6	5	8	3	5	4	7	2	4	3	6	1
4	14	19	7	3	8	18	6	17	7	12	0	11	1	6	14	10	0	5	13
Yaxkin	Zac	Zec	Xul	Cumhu	Zotz'	Pax	Kayab	Yax	Muan	Ch'en	Yax	Zip	Mol	Uo	Uo	Kankin	Uayeb	Mac	Mac

From pages 46–50 of the *Dresden Codex*.

and 8 days, the periods of visibility and invisibility. From this they
derived an almanac for Venus [146].

The table starts with the date 3 Cib 4 Yaxkin (the 4 Yaxkin is found in
the bottom two lines). This is a date on which Venus disappears. It
reappears 90 days later, on 2 Cimi 14 Zac. 250 days later, 5 Zib 19 Zec, it
disappears again, to reappear again 8 days later, 13 Kan 7 Xul. A final
236 days of invisibility take us through one synodic period. Five such
periods take us right across the chart and we continue with the second
line, 11 Cib 4 Yaxkin. The reason why the 4 Yaxkin repeats is that the
number of days in five synodic periods is a multiple of 20 and also of 365.
After 65 periods (13 lines of the table) we come back to 3 Cib 4 Yaxkin,
and everything repeats.

So far so good, but the average synodic period of Venus is not quite
584 days: it is 583.92 days. Therefore the almanac needs correction; and
the Mayas corrected by 20 days every 240 periods as follows:

After	61 periods subtract	4 days
After	61 periods subtract	4 days
After	61 periods subtract	4 days
After	57 periods subtract	8 days
Total:	240	20

This does give an average synodic period of 583.92 days. The reason why
they corrected like this instead of just subtracting one day every twelve
periods is to make the almanac simpler. For example, 61 periods after our
first date, 3 Cib 4 Yaxkin, we end up 61 × 584 − 4 days later, i.e., 35,620
days later. This is a multiple of 13 and of 20, so the date we have reached
is also a 3 Cib day. If we divide 35,620 by 365 we get a remainder 215, so
to get the second part of the date we count 215 days on from 4 Yaxkin.
215 days = 10 uinal + 15 days, so this brings us to 19 Kayab. The next
synodic period, with this correction, starts at 3 Cib 19 Kayab. Therefore
we need not alter the top part of the table; we need only replace the
bottom two lines by lines starting 19 Kayab. And indeed the almanac
does, lower down, have such a pair of lines. They read:

19	4	14	2	...	3
Kayab	Zotz	Pax	Kayab		Xul.

The same applies to the other correction, because 57 × 584 − 8 is also a
multiple of both 13 and 20.

The four parts into which the Mayas divided the synodic period of
Venus are not accurate. The true average times are 263, 50, 263, 8 days,
not 236, 90, 250, 8; the value of 90 instead of 50 for the first period of

invisibility is particularly far out. Anthony F. Aveni has suggested that the figures in the almanac are not purely astronomical but a combination of astronomy and some kind of ritual numerology [147].

Eclipse Table

Pages 51 to 58 of the *Dresden Codex* contain 69 interesting columns. They start on the top halves of pages 53 to 58 and finish on the bottom halves of pages 51 to 58.

Figure 9.2(a) shows the first eighteen columns (the top halves of pages 53 to 55). Levels A and C are numbers and the three rows at level B are the first halves of calendar-round dates. Figure 9.2(b) shows the first six columns transliterated. If we convert the numbers to decimals and replace each date by its position in the 260-day cycle (1 Imix = 1, 2 Ik = 2, 1 Ix = 14, and so on) the table starts and ends as in Figure 9.2(c).

The numbers at level C are mostly 177, but eight of them are 148; after each of these a picture is inserted—two can be seen in Figure 9.2(a). It is

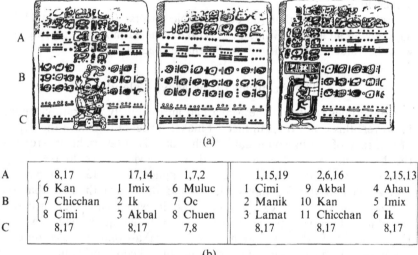

(a)

A	8,17	17,14	1,7,2		1,15,19	2,6,16	2,15,13
	6 Kan	1 Imix	6 Muluc		1 Cimi	9 Akbal	4 Ahau
B	7 Chicchan	2 Ik	7 Oc		2 Manik	10 Kan	5 Imix
	8 Cimi	3 Akbal	8 Chuen		3 Lamat	11 Chicchan	6 Ik
C	8,17	8,17	7,8		8,17	8,17	8,17

(b)

A	Running total	177	354	502	679	856	1033	1211	...	11,604	11,781	11,958
B	Day numbers	84	1	149	66	243	160	78		72	249	166
		85	2	150	67	244	161	79		73	250	167
		86	3	151	68	245	162	80		74	251	168
C		177	177	148	177	177	177	177	...	177	177	177

(c)

FIGURE 9.2.

these numbers that make it clear that we have an eclipse table. As we saw on page 18 the shortest intervals between successive eclipses are 5 months and 6 months, i.e., on average, 177 days and 148 days, the 148-day interval being rarer than the 177-day interval.

The numbers at level A are a running total, though on six occasions the total increases by an extra 1: by 178 instead of by 177. We do not know how the distribution of the 148-day intervals was decided, but it does make the tabled dates occur very unevenly in the 260-day cycle. They are crammed into three sectors: 57 to 91, 144 to 178, and 231 to 1 (1 is of course equivalent to 261 in the cycle). Not only that, but any two dates in the same sector are an even number of sectors apart. Therefore the table is based on a double cycle, and the 69 entries are squeezed into three sectors evenly spaced round the double cycle, namely, 144 to 178, 317 to 351, and 490 to 591, covering a total of only 109 of the 520 possible positions. There is a sound reason for this. An eclipse cannot occur unless the sun is at a node of the moon's orbit, and the time taken for the sun to go from one node to the next is always very close to 173 days; 173 is one-third of 520. All this looks again like a combination of astronomy and numerology.

It is worth noting that the eclipse table covers 405 months in 11,958 days, giving an average value of 2,391.6 days in 81 months, agreeing well with the value 2,392 deduced for Palenque (page 199).

The Accuracy of the Maya Calendar

We sometimes read that the Maya had an extremely accurate calendar. However, this is not true; in fact , it makes no sense to talk of the degree of accuracy of a long-count date. Such a date is either right or wrong: if the Maya missed a day or counted a day twice it would be wrong; otherwise it would be right. The same applies to the 13-by-20 cycle. It also applies to the 365-day cycle (unless the Maya thought that the astronomical year was exactly 365 days, in which case their accuracy would have been very poor). This was all explained quite clearly by John E. Teeple, a leading researcher into Maya astronomy, according to whom the Maya calendar "made no attempt to keep itself adjusted to the seasons, as our calendar does by inserting leap-year days. It was simply an arbitrary and orderly succession of days and months in regular order, going on for ever without regard to any natural phenomena. We can infer that they knew the length of the year to be 365 days or better, but beyond this the term 'accuracy' is meaningless in connection with the Maya calendar, just as meaningless as to speak of the accuracy of our seven-day week."

It is ironical that Teeple should have been so definite about this, because the misconception about the Maya calendar can be traced back to his writings [148]. He thought that the Maya knew the length of the year quite accurately, and in fact made it 365.2420 days. I will explain below how he got this figure. Modern reference books give the figure 365.2422 days. The average length of the year in our (Gregorian) calendar is 365.2425 days. Consequently (if Teeple is right) we have, for the length of the year in days:

	From the calendar	Best estimate
Maya	365	365.2420
Modern	365.2425	365.2422

Unfortunately, Teeple compared the Maya's best estimate—not their calendar-value—with the modern calendar-value, and wrote "the Maya computation was far better than that of the Julian calendar, which was used in our country until after A.D. 1700, and, in fact, is almost identical with our present Gregorian calendar."

Teeple deduced the Mayas' value for the length of the year as follows. At Copan, Stela A has a lengthy inscription containing three dates (all about A.D. 700):

   (i) 3,899 tun 160 days, 12 Ahau, 18 Cumhu
  (ii) 3,900 tun   0 days,   4 Ahau, 13 Yax
 (iii) 3,899 tun 100 days,   4 Ahau, 18 Muan.

The first part of date (ii) was actually written

9 baktuns 15 katuns 0 tuns 0 uinals 0 days.

The three zeros show that it is the end of a katun, and the Maya were particularly interested in ends of katuns, to judge by the large number of such dates that have been found. Because the year is not exactly 365 days long, the day 13 Yax in A.D. 700 will not occur at the same season of the year as 13 Yax at the beginning of the Maya calendar (baktun zero, about 3000 B.C.). Teeple suggested that the stela answers the question "of what day in the epoch of baktun zero katun zero is date (ii) the true astronomical anniversary?", the answer being 18 Cumhu: date (i) is the full Maya date of the last 18 Cumhu before date (ii). He also suggested that date (iii) displays the basis for the calculations; he interpreted it as saying that the sun is in the same part of its annual cycle as it was at the end of the previous katun, and the moon is in the same phase. (Although we cannot decipher the inscription there is some evidence that the sun and the moon are involved.) Date (iii) is 6,940 days after the end of the previous katun so Teeple's suggestion would mean that 6,940 days is both an exact number of years and an exact number of months. The number of years can only be 19; the number of months, 235. Thus

$$6,940 \text{ days} = 235 \text{ months} = 19 \text{ years}.$$

This by itself would give

$$1 \text{ year} = 6940 \div 19 \text{ days}$$
$$= 365.2632 \text{ days}.$$

This is not very accurate—not accurate enough to give the correct answer 18 Cumhu. But Teeple ignored the 6,940 days and combined the formula

$$235 \text{ months} = 19 \text{ years}$$

with the formula

$$149 \text{ months} = 4,400 \text{ days}.$$

This is the formula at Copan that corresponds to the formula 81 months = 2,392 days at Palenque (page 199). Teeple quoted it without showing how he computed it. It is less accurate than the Palenque formula. The two formulas between them give

$$1 \text{ year} = (235/19) \times (4,400/149) \text{ days}$$
$$= 365.2420 \text{ days}.$$

Two errors have cancelled, giving an undeservedly accurate result.

CHAPTER 10

The European Renaissance

Copernicus

We now return to Europe and look at Copernicus (A.D. 1473–1543). His treatise *De revolutionibus orbium coelestium* [On the revolutions of the heavenly spheres], published in the year of his death, is as substantial as the *Almagest*, but I will not describe it in such detail. What everyone "knows" about Copernicus is that he made the sun stand still: he considered the sun to be fixed at the center of the universe, the planets to be circling round it, and the earth to move just like any other planet. But this is misleading, if not actually false: Copernicus did not place the sun *at* the center but *near* the center of the various orbits, and the earth does not move quite like the other planets (it does not have an epicycle (see page 208)).

The preface to *De revolutionibus* was probably written not by Copernicus himself but by the theologian Andreas Osiander, who took a friendly interest in the book. He pointed out that the rôle of an astronomer is to construct a theory from which astronomical movements could be calculated; the theory need not be true, it need only fit the observations, and neither an astronomer nor a philosopher will find certainty unless it is "divinely revealed." This enabled people who could not stomach a moving earth for theological reasons to teach Copernicus's system as a mathematically convenient but physically erroneous hypothesis; and indeed the book was not banned until, some sixty years later, Galileo pushed it hard and the church reacted (some would say overreacted).

De revolutionibus is closely modeled on the *Almagest*: a tribute to the authority of that work (well over a thousand years old by 1543). It starts by stating that the universe and the earth are spherical and that heavenly motions are made up of regular circular motions. Then comes the main point in which Copernicus disagreed with Ptolemy. Does the earth rotate? Ptolemy had argued that if the earth rotated everything would be swept toward the west. Copernicus countered this by saying that anyone who maintains that the earth rotates will postulate that the rotation is "natural." Here he was maintaining Aristotle's distinction between natural motion (which is circular for celestial bodies but rectilinear for earthly bodies) and unnatural motion. He assumed that Ptolemy's reason for thinking that the earth's rotation would have a disastrous effect was that such motion is "violent" (which in the Aristotelian philosophy of motion means simply "unnatural"). Copernicus argued that if Ptolemy was worried by a rotating earth he should have been even more worried by a rotating sky, whose "immensity" would "increase with the increase in movement" and so become infinite. He quoted a supposed axiom of physics that "nothing infinite can be moved" and concluded that the heavens would come to rest. He added that immobility is nobler and more godlike than instability and so should belong to the sky rather than to the earth.

These arguments will strike the modern reader as, to say the least, implausible. The statement, which we read all too often, that Copernicus proved that the earth moves round the sun is simply not true. Nevertheless, sun-at-rest astronomy completely superseded the earth-at-rest astronomies described so far. Why?

In a quite trivial sense, any geostatic theory is equivalent to a heliostatic theory and vice versa. If we make a scale model in a flat box of a geostatic solar system, driven perhaps by an electric motor, place it on a smooth table, and hold the sun still, allowing the box to slide around on the table, we obtain a heliostatic system. But if the geostatic system is Ptolemy's or Āryabhaṭa's, the resulting heliostatic system will not make much sense. If, on the other hand, we start with the geostatic system described on page 175 the resulting heliostatic system does make sense. Such a geostatic system was, in fact, proposed by Tycho Brahe and we might say (details aside) that holding the sun still in Tycho's system gives us Copernicus, while holding the earth still in Copernicus's system gives us Tycho.

Modern astronomy is not heliostatic. Although it is convenient, when dealing with the solar system, to consider motion relative to the sun, not to the earth, the sun is not at rest but moves round the center of our Galaxy. Is the center of the Galaxy at rest? The galaxies are certainly moving relative to each other and the question of which, if any, is "really" at rest is meaningless. All motion is relative. In fact, the difference between geostatic and heliostatic systems is not of great technical astronomical importance. Its importance, if any, is theological and philosophical.

Even Galileo had trouble in finding valid reasons for believing that the earth, not the sun, moved. He was reduced to citing the tides, which he thought were caused by the motion of the earth, like water sloshing round in a tub carried on a moving cart [148a].

Not only were there difficulties in finding reasons for believing that the earth moved but there were at least two reasons for not believing this. One was the argument that motion of the earth in orbit would have the same effect on comets as on planets and would make them retrogress; the other was that this motion would cause parallax in the directions of the stars, observed from one position now and from a different position six months ago. No such parallax was detected. (In fact, comets move too fast to be made to retrogress, and annual parallax is too small to be detectable by the instruments of the time.)

If heliostatic astronomy has no intrinsic technical advantages, if thoroughly modern astronomy is not heliostatic, and if the arguments adduced by Copernicus were unconvincing, why were the early geostatic astronomies, Ptolemy's in particular, superseded? The great advantage of Copernicus's system (which, admittedly, it shares with Tycho's) is that it put the motions of the planets together coherently, as opposed to the *Almagest*, which kept them separate, or the *Hypotheseis* or the *Āryabhaṭīya*, which put them together incorrectly. In particular, it explained why the planets' retrogressions all took place while the planets were in line with the sun; for earlier theories these were five unexplained coincidences. The important thing about Copernicus's theory is not that it is heliostatic but that it is heliocentric. A system can be heliocentric without being heliostatic—Tycho's system was. And, logically, a system can be heliostatic without being heliocentric, though no such system has ever been advocated.

The advantages of a coherent heliocentric system were very compelling, especially after it was realized that the planes of the planets' orbits pass through the sun, not the earth, enormously simplifying the treatment of the planets' latitudes. But the main reason is that in the seventeenth century, Newtonian mechanics, in which there is such a thing as absolute motion, was able to explain on a heliostatic basis the motions of the planets (with the exception of the anomalous precession of Mercury, which had to wait for the general theory of Relativity) and could also explain Coriolis forces and the motion of Foucault's pendulum, which would have been utterly mysterious if the earth were at rest. It was in the period between Newton's time and the discovery of the theory of Relativity (which showed that there is no such thing as absolute motion) that the "Copernican revolution" was achieved and heliostatic astronomy swept earlier theories aside.

Copernicus continued by describing in general terms how the earth moves in his theory—it moves round the sun, at the same time spinning on its own axis, which always points in the same direction in space—and, still following the pattern of the *Almagest*, he finished Book 1 with geometrical theorems and a table of chords.

Copernicus's treatment of precession (Book 2) is complicated because he thought he detected a change in the rate at which the equinox-point moves. His theory of the motion of the earth is the same as Hipparchus's theory of the motion of the sun. He found the basic parameters from the same kind of data: the length of the year and the lengths of two seasons. He had, of course, over a thousand years of observations after Ptolemy at his disposal. In his calculations he used sidereal years, not tropical years. His results are:

Eccentric-quotient	116/3600
Longitude of apogee	96°40′.

When Copernicus turned his attention to the moon he pointed out that Ptolemy's theory (page 148) gave incorrect values to the greatest and least distances of the moon from the earth, and adopted the same device as al-Shāṭir (page 193). The ratios of the radii of the extra epicycle, the main epicycle and the deferent turned out to be 24 to 110 to 1,000. Copernicus then repeated Ptolemy's calculation of the distance of the sun, getting a slightly worse result: 1,179 earth's radii.

Copernicus made each planet (except for Mercury) move on a small epicycle whose center C moves on a deferent whose center D is not located at the center X of the earth's orbit. Let us call this an *epicyclet*; it does not play the same rôle as Ptolemy's epicycle—its purpose is to account for the same small irregularities as Ptolemy's equant. The radius of the epicyclet is $\frac{1}{3}DX$: this fact is part of the theory, it is not calculated from observations. The epicyclet moves round the deferent in the sidereal period of the planet, and the planet moves round the epicyclet at a speed that keeps the angles PCD and CDA equal (see Figure 10.1). Copernicus calculated his basic parameters from observations, just as Ptolemy did,

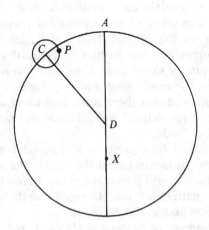

FIGURE 10.1.

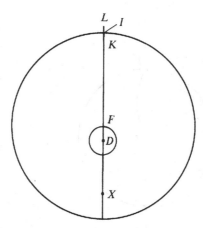

FIGURE 10.2.

but first he showed that the basic parameters which Ptolemy found, when applied to Copernicus's theory, agree with the observations which Ptolemy used. To do this, he took DX to be three-quarters of Ptolemy's eccentric-distance, and therefore CP to be one-quarter. We might say that he took Ptolemy's eccentric-distance and shared it between his own eccentric-distance and his epicyclet.

Like Ptolemy, Copernicus needed a more complicated theory for Mercury (see Figure 10.2). X is the center of the earth's orbit. The radius of the small circle center D is $\frac{1}{3}DX$. The point F moves round this circle in the same direction as the earth at two revolutions per year. The point I moves round a circle center F in the sidereal period of Mercury. $LIKF$ is a straight line, $LK = DF$, and I is the midpoint of LK. Mercury moves from K to L and back again twice a year, starting at K when $XDFI$ are in line. This straight-line motion can be produced by combining two circular motions as Copernicus had explained earlier when he used it to produce the variation he thought he had found in precession. If (in Figure 10.3(a)) J moves round the circle center I and at the same time P moves in the opposite direction round a circle center J in such a way that the rate of revolution of PJ relative to JI is twice that of JI, starting with IJP in line, then P moves along this line. The reason is that if JP cuts the line through the initial position of IJP in P^* (Figure 10.3(b)) then $P^*JI = 180° - 2P^*IJ$ and so $JP^*I = JIP$. Then $JP^* = JI = JP$ and so $P = P^*$. Therefore P is on the line in question. (The last two links in Naṣir al-Din's linkage, Figure 8.1, have precisely the same effect.)

In the broad history of science, of its relations with theology, and of its place in civilized thought, Copernicus stands at the start of a new era. But in the narrower technical history of astronomy he is the last of the old rather than the first of the new. The shapes of the orbits in his theory are

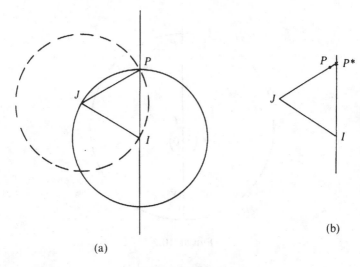

(a)

(b)

FIGURE 10.3.

only a byproduct of the theory and may or may not be the same shape as the orbits that the earth, moon, and planets actually follow. The new era did not get under way until someone found a way of checking on the shapes of the orbits and coordinating distances with angles. That man was Kepler.

Tycho Brahe

Before we look at Kepler's work, however, we must look at Tycho Brahe's (see Figure 10.4). Brahe (A.D. 1546–1601) not only refined the technique of observation far beyond previous observers (to an accuracy of between 30″ and 50″ according to the instrument used) [149] but made important theoretical advances too.

Brahe was convinced that the earth is at rest. He had a lively correspondence with Christopher Rothman, chief astronomer to the Landgrave of Hesse, in which he argued that if the earth were moving a stone dropped from a tower would not fall vertically [150]. Rothman made what is, in fact, the modern reply: that both the stone and the tower take part in the motion of the earth before and during the fall. To this Brahe replied that he could not believe that a body could have two motions at once; one would interfere with the other

Some modern commentators pour undue scorn on this argument. J.L.E. Dreyer was particularly scathing, finding it "very curious" that Brahe did not make the "simple experiment" of dropping a stone from the mast of a "swiftly-moving vessel" [151]. Such commentators must

FIGURE 10.4. Danish postage stamp. (Collection E.C. Krupp, Griffith Observatory.)

be the veriest land-lubbers. Where could Brahe find a "swiftly-moving vessel" that also moves smoothly? If an astronomer did manage to clamber to the yard-arm of a ship moving in a stiff breeze, his stone would be as likely to fall over the side as anywhere [152]. Pierre Gassendi did indeed claim, about A.D. 1640, to have made this experiment using a trireme moving at 16 knots, but anyone who has witnessed the jerky motion of a hard-rowed vessel under the synchronized strokes of the oars could be excused for being sceptical [155].

The clinching argument for Brahe seems to have been his belief that the motion of the earth would have the same effect on comets as on planets and make their motion retrograde when they are in opposition to the sun. (He had investigated comets and found that they were more distant than the moon and consequently belonged properly to astronomy, not meteorology as earlier scientists had thought.) Brahe therefore rejected Copernicus's theory and proposed his own—the theory we mentioned on page 175 in which the planets move round the sun while the sun moves round the earth. As I remarked earlier, this was not possible until after his research into comets had shattered his belief in the solidity of the spheres carrying the heavenly bodies.

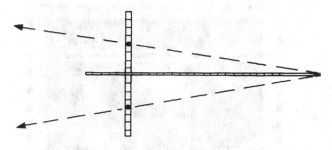

FIGURE 10.5.

Brahe's interest in astronomy was aroused when, at the age of four-teen, he saw an eclipse of the sun, and was impressed by the fact that such things could be foretold. However, when a conjunction of Saturn with Jupiter took place three years later, current predictions, using the Prutenic tables, were several days out, and this stimulated Brahe to investigate the motions of heavenly bodies for himself. His first instru-ment was a simple cross-staff (see Figure 10.5). A graduated bar, some-thing like a yard-stick, carried a backsight at one end. Another graduated bar at right angles to it could slide along it, and carried two foresights. The observer set up the instrument so that he could see an object through the backsight and one foresight, and another object through the backsight and the other foresight. The angle subtended by the objects at the observer's eye could be calculated from the scale readings. Even at this early age Brahe showed signs of the genius to come. He noticed that the marked graduations were not quite accurate, and he constructed a table of corrections.

In 1569 Brahe started to build accurate instruments, starting with an oak quadrant with a radius of 6 metres—large enough for every minute to be marked. This quadrant could be turned about a vertical axis. He also built "sextants"—nothing to do with the modern navigational instrument of that name, but in effect a moderate-sized quadrant cut down from a quarter of a circle to one-sixth of a circle (whence the name) and attached to a stand by a universal joint, so that it could be turned in any direction (see Figure 10.6).

In 1572 one of the most impressive events in astronomical history occurred—a new star appeared in the sky. Nowadays, we should call such a star a supernova. Only eight supernovae are known to have appeared, in A.D. 185, 385, 393, 1006, 1054, 1181, 1572, and 1604, and of these only the ones in A.D. 1006, 1572, and 1604 were noticed in the Western world. Not only was the star an impressive sight—to begin with, as bright as Venus at its brightest, and therefore much brighter than first magnitude—but it even had philosophical and theological implications: medieval theo-logians had sharpened Aristotle's theories to the point where they asserted

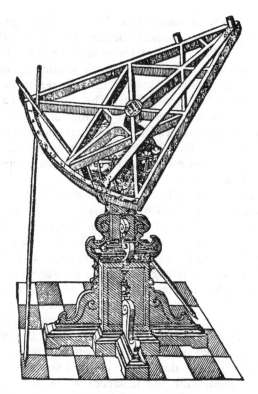

FIGURE 10.6. Tycho's great sextant.

that the heavens were perfect. Perfection implies completeness, which means that no new stars could appear. The supernova lasted for eighteen months, and as it faded its color changed to yellow, then to red, and finally to livid. Brahe investigated it thoroughly, repeatedly measuring its angular distances from the nearest stars (those in the constellation Cassiopeia). He published a book about it (abbreviated title *De Nova Stella*) in 1573, although at first he was reluctant to do so, because at that date it was considered improper for noblemen to write books, and Brahe was indeed of noble descent.

Brahe's account of the new star ended with a statement of its astrological implications, namely, that the star's influence would begin to be felt in 1592 and would last until 1632, and that religions full of pomp and splendor would disappear. As Gustaf Adolf, the great Protestant champion, was born in 1594 and died in 1632, many people were impressed by this prediction. Brahe seems to have believed in astrology in spite of a failure early on, when he said that an eclipse of the moon in October 1566 foretold the death of the Sultan of Turkey (who was eighty years old). It turned out that the Sultan had died in September, but news traveled slowly in those days. In 1574 Brahe gave a lecture in Copenhagen on the mathematical sciences. In it he described three uses of astronomy: it enables us to determine the time, it exalts the mind from earthly trivialities, and it enables us to draw conclusions about human fate from celestial motions. He also tried to refute logically some of the common objections to astrology. The objection that many people with greatly different horoscopes die together in a flood or a plague he countered by saying that prudent astrologers make reservations about natural calamities. To the objection that many people with the same horoscope have vastly different fates he replied that diversity in education and mode of life account for the divergences.

In 1576, King Frederick II of Denmark gave Brahe what is probably the most magnificent gift a king has ever given to an astronomer—a whole island, Hven (or Hveen), just east of Copenhagen. It is now part of Sweden, and its Swedish name is Ven. On this island Brahe built an observatory/palace which he called Uraniborg, where he set up many accurate instruments, including the *armilla aequatoria maxima* of Figure 1.21, and worked diligently there until 1597, when royal funding stopped.

Brahe measured the angle between the ecliptic and the celestial equator and found it to be 23°31′30″. The correct value for that date was 23°30′. Copernicus had found 23°28′: at his date the correct value was 23°31′. Brahe was twice as accurate as Copernicus, but too high instead of too low. Brahe's error is due to the poorly known distance of the sun (Brahe did not re-measure this distance) which gave a parallax of 3′, whereas the sun's parallax is actually quite negligible. The improvement over Copernicus is due to the fact that Brahe took account of refraction; he investigated refraction quite thoroughly, producing a table of refraction for the sun, and one for the stars.

He also located the apogee of the sun's orbit, and found that it changed by 45″ per year. The true value was 61″; Copernicus had found 24″. He measured the tropical year and found it to be 365 days, 5 hours, 48 minutes, 45 seconds. This result is about one second too small.

From 1582 onward Brahe made numerous observations of the moon. His vast file of accurate observations at all points of the orbit, not just at full moon, half moon, and other special positions, was his great advantage over Copernicus and other earlier astronomers. He found an anomaly in the moon's motion (i.e., a periodic deviation from previous theories) that reached a maximum of $40\frac{1}{2}''$ when the moon is one-quarter or three-quarters full. He produced a theory of the moon's motion which allowed for this, the moon moving round an epicycle which itself moved round an epicycle which moved round a moving deferent. He then found a deviation from this theory, with a period of half a month, which he accounted for by letting the speed of the second epicycle along the deferent vary. He also found a small deviation with a period of one year. His assistant, Christian Severin, accounted for this, not by an extra geometrical complication, but by incorporating a correction in the calculations. Brahe also found that the angle between the moon's orbit and the ecliptic is not constant, but oscillates between 4°58'30" and 5°17'30". Finally, he found that the rate of retrogression of the nodes varies. Putting this all together, Brahe ended up with a vastly improved theory of the moon's motion.

Brahe's most important publication was *Astronomiae Instauratae Progymnasmata*, a three-volume work, of which only the second volume was ever completed. This second volume was mainly about a comet which had appeared in 1577. As a byproduct of his work on this comet, Brahe found accurate positions for the stars used in the observations. It is in this book that he sketched his earth-at-rest solar system (see page 175); he promised a fuller description later, but none ever appeared. The first volume contained over 900 pages. It dealt with Brahe's theory of the sun's motion, his work on precession (he found that the rate of precession did not change, although Copernicus had thought that it did—see page 207) a catalogue of stars, and an account of the supernova of 1572.

Brahe's other substantial publication was *Astronomiae Instauratae Mechanica*, 1598. It consisted chiefly of a description of his instruments and of the Uraniborg observatory, but it also included a biography, an account of some of his discoveries about the sun and the moon, some observations of comets and planets, and a painstakingly compiled extremely accurate catalogue of the positions of a thousand stars. (This catalogue was also issued separately.)

One feature of Brahe's instruments was the use of *transversals*, the dotted lines (each consisting of ten dots) shown in Figure 10.7. It would not be possible to mark ten subdivisions on the scale between the divisions shown, but we can see easily that the pointer is showing 2.4 by noting how far up it crosses the transversal. (This device had been invented by Johann Hommel, 1518–1562.)

Brahe also improved the sights on the instruments. Instead of a hole in a plate, Brahe's backsight consisted of four slits along the four sides of a square (Figure 10.8). The foresight consisted of a square of the same size, and the observer lines a star up with the top edge of the foresight through the top slit, with the right-hand edge of the foresight through the right-hand slit, and so on. The slits could be narrowed for greater accuracy when a bright star was being viewed, and widened for better visibility if the star being viewed was faint.

Brahe did not favor using clocks for timing his observations: the time could be found more accurately by astronomical means.

In 1598, the year after the king of Denmark ceased to support Brahe at Hven, he was invited to Prague by the emperor, Rudolph II. In 1601 he was joined there by Kepler, who was in need of accurate observations for his own researches. Brahe had written an interesting letter to Kepler in which he pointed out a discrepancy in the directions of Mars from the

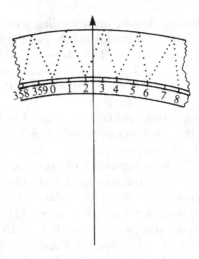

FIGURE 10.7.

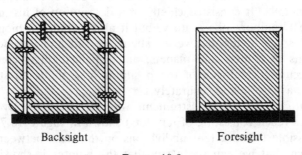

Backsight   Foresight

FIGURE 10.8.

earth at different times: two angles which should be equal turned out not
to be equal. If this were thought of from an earth-goes-round-the-sun
viewpoint, which Kepler preferred, it would mean that the earth's orbit
expanded and shrank periodically. (From a sun-goes-round-the-earth
viewpoint, it would mean that the sun's orbit expanded and shrank.)

Let us explain this using the sun-at-rest point of view. If D is the center
of the earth's orbit and S the sun (Figure 10.9(a)), then the direction of
SD is known: it is the direction of aphelion, T_1, the point on the orbit
furthest from the sun. The times when the direction of Mars is perpendicular to SD are known. So are the times when the earth is at T_1 and T_2
(perihelion). It is at T_3 half-way between these two times; this is because
in all theories known at this time (Ptolemy's, Copernicus's, and Brahe's,
in particular) the movement of the earth relative to the sun is uniform
round D. Observation gives the directions T_1M, T_2M, and T_3M, and so
yields the angles T_1MT_3 and T_2MT_3. These should be equal. In fact,

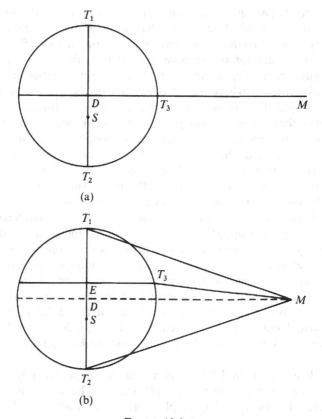

(a)

(b)

FIGURE 10.9.

T_2MT_3 turned out to be appreciably larger than T_1MT_3. A change in the size of the earth's orbit would account for this: if the orbit is smaller when the earth is at aphelion, T_1 is drawn closer to D and the angle T_1MT_3 shrinks. But no one wanted to assume that the orbit changed: that would upset every theory devised so far. Kepler's explanation was that the earth does not move uniformly about the center of its orbit but, like every other planet, has an equant (E in Figure 10.9(b), which shows how this suggestion makes T_2MT_3 larger than T_1MT_3).

Kepler

Kepler's main work in astronomy is contained in a vast volume of some 400 folio-sized pages, *Astronomia nova aitologetos*, published in 1609. It is a very different kind of book from the *Almagest* or *De revolutionibus*, in which Ptolemy and Copernicus presented formally the results, and only the results, of their labors. Kepler wrote informally, describing his motives, his hopes, his false starts, his frustrations and—finally—his triumph. It is not his first book: that was *Prodromos dissertationem cosmographicarum continens mysterium cosmographicum*, 1596, in which he tried to account for the relative sizes of the orbits of the planets by fitting regular polyhedra (a cube, a tetrahedron, etc.) between them.

This in itself led nowhere, but in the course of his investigations Kepler became convinced (a conviction that turned out to be fundamental to his later work) that the important point in the solar system, on which the geometry of the various orbits should be based, is the sun—not, as Copernicus suggested, the center of the earth's orbit. (The center of the earth's orbit is sometimes called the "mean sun," because it plays the same part in Copernicus's theory as the mean sun does in Ptolemy's: the line joining it to the earth rotates at a constant rate.)

When Kepler arrived at Prague, Brahe's partner, Christian Severin, had been observing the oppositions of Mars, and he and Brahe had worked out a theory of its motion, along the same lines as Copernicus, using an eccentric deferent and an epicyclet, but not adhering to Copernicus's dogma that the radius of the epicyclet is one-third of the eccentric-distance. In fact, they made the radius of the epicyclet 0.2016 and the eccentric-distance 0.1638 (where the radius of the deferent is 1). This theory turned out to be very accurate for longitudes when Mars is in opposition; it is not so good for other longitudes and quite poor for latitudes.

When Brahe died in 1601, Kepler continued: he remarked that had Severin been studying some other planet, then he would have investigated that other planet [153]. Kepler was lucky—had he studied Venus, Jupiter, or Saturn his methods would not have been effective because their orbits were too close to being circular; had he studied Mercury he would have found that available observations were not good enough.

Kepler's attack on Mars is described in full in *Astronomia nova*. Part I, which I will not describe in any detail, is introductory and historical. Part II starts with a painstaking piece of geometry [Chapters 5 and 6] showing that if the center of a planet's orbit were incorrectly located by a distance equal to that between the sun and the mean sun, the equant being correctly located, the longitudes in opposition would still be accurate to within 5', but other longitudes would be wrong and so would latitudes. This is just what happened under Severin and Brahe's theory; and presumably this fact encouraged Kepler to switch from the mean sun to the true sun.

Brahe's table of longitudes of Mars when in opposition looked like this:

Date and time	Longitude in orbit	Longitude in ecliptic	Mean sun
1580 xi 17 0940	66°50′10″	66°46′10″	246°48′32″
1582 xii 28 1216	106°51′30″	106°46′10″	286°50′58″
1585 i 31 1935	141° 9′50″	141°10′26″	321°10′13″
1587 iii 7 1722	175° 5′10″	175°10′20″	355° 5′57″
1589 iv 15 1334	213°54′35″	213°58′10″	33°53′32″
1591 vi 8 1625	266°42′ 0″	266°32′ 0″	86°45′24″
1593 viii 24 0213	342°35′ 0″	342°43′45″	312°34′36″
1595 x 29 2122	47°56′ 5″	47°56′15″	227°56′17″
1597 xii 13 1335	92°34′ 0″	92°28′ 0″	272°28′51″
1600 i 19 0940	128°18′45″	128°18′ 0″	308°18′43″

(Incidentally, this table gives some idea of the irregularity of Mars's motion. The interval between the first and second oppositions is 761 days; between the sixth and seventh it is as much as 808 days.)

Kepler's first move was to calculate precisely the position of the mean sun at each of the times listed, in order to see how accurately Mars was in opposition. The greatest error was 13′24″ (the 1591 opposition). He then compared these positions with the longitude of Mars in its orbit (instead of along the ecliptic) and found much smaller errors; the biggest was 3′24″, and eight of the rest were less than 1′40″. This inspired him to devote a whole chapter [Chapter 9] to the question of exactly when opposition occurs.

In our diagrams S will denote the sun, T the earth, M Mars, D the center of Mars's orbit, and E Mars's equant. Figure 10.10 shows a section through S, M, and T. C is the apparent position of Mars on the celestial sphere (whose center is T): that is, the apparent position of Mars to an observer on earth. F is Mars's "real" position: the apparent position of Mars to an observer on the sun. Thus the "real" path of Mars lies between its apparent path and the ecliptic.

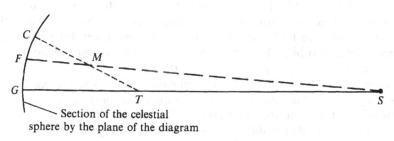

Section of the celestial sphere by the plane of the diagram

FIGURE 10.10.

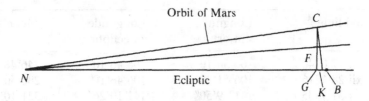

FIGURE 10.11.

Figure 10.11 shows part of the orbit of Mars as seen from the earth; C and F are still its apparent and real positions, N is a node (a point where the latitude of Mars is zero) and $NGKB$ is the ecliptic. The ancients thought that opposition took place when the sun was opposite G, where CG is perpendicular to the ecliptic. Brahe and Severin evidently thought that it took place when the sun was opposite B, where $NB = NC$. Kepler maintained that it took place when the sun is opposite K, where CK is perpendicular to the "real" orbit NF.

In the next two chapters, Kepler investigated the parallax of Mars and came to the conclusions:

(i) it is difficult to deal with; and
(ii) it is not important because Mars is mostly observed high in the sky, in which position its parallax is small.

(We know today that this was an unnecessary worry. But in Kepler's time astronomers badly underestimated the size of the solar system. Kepler himself stated [end of Chapter 11] that the distance of the sun is between 700 and 2,000 earth's radii. The true figure is about 24,000.)

Kepler described graphically some of his difficulties. On a cold windy night he could not close the iron clamps of his apparatus with his bare hands; but with gloves on he could not adjust the apparatus as precisely as he needed.

The Latitude of Mars

The preliminaries are over; the real battle starts. Kepler began with the latitude [Chapter 12]. First he found four observations when Mars was crossing the ecliptic from north to south. The intervals between the observations were multiples of 687 days, the time taken for Mars to go once round its orbit. This shows that the instants at which Mars is at a node are not affected by the motion of the earth.

From Copernicus's theory for the longitude of Mars Kepler calculated the longitude of the node (as seen from the mean sun) and found that it was 46°48'. He found the other node to have longitude $225°44\frac{1}{2}'$. Because these two figures do not differ by 180°, the line joining the nodes does not pass through the mean sun. By the time Kepler was writing this he had

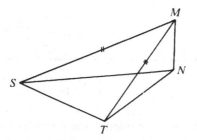

FIGURE 10.12.

completed his research, so he knew that the line joining the nodes passes through the real sun, not the mean sun. However, he contented himself here with the remark that he would return to this detail later (which he did in Chapter 61).

Next he looked through the files for observations of Mars made when $SM = TM$ and SM is at right angles to the line joining the nodes (Figure 10.12). If MN is the perpendicular from M to the ecliptic, then:

(i) MSN is the angle between the plane of Mars's orbit and the ecliptic; and

(ii) $MSN = MTN$, the observed latitude of Mars.

From half a dozen observations when Mars is in or near this configuration, he found that Mars's orbit is inclined at 1°50′ to the ecliptic.

He confirmed this by a second method: he found a few observations made when T is on the line joining the nodes and STM is a right angle (Figure 10.13). Then the angle MTN is the inclination required. This method has the advantage that it does not depend on any particular theory for the distances from Mars and the earth to the sun.

Finally, Kepler checked yet a third time, using observations made at opposition: it is an easy piece of trigonometry to calculate MSN (Figure 10.14) from MTN, which is observed, and the ratio of ST to TM which is given by Brahe's theory of Mars's motion.

The inclination of Mars's orbit to the ecliptic turned out to be 1°50′ each time. This shows that the plane of the orbit is fixed: it does not tilt as

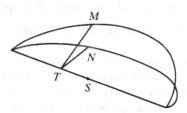

FIGURE 10.13.

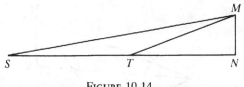

FIGURE 10.14.

Copernicus thought. Some modern commentators have gone astray here; they suggest that the reason why Copernicus found the orbit tilting was that he made the plane of the orbit go through the mean sun, not the real sun. But Kepler's calculations so far are based on the mean sun, as he says clearly at the end of Chapter 12. His own explanation of Copernicus's mistake was that Copernicus followed Ptolemy too closely.

The First Theory of Mars

When Kepler turned his attention to the longitudes [Chapter 15] he did switch to the real sun, and his first task was to modify Brahe's table to give true oppositions instead of mean oppositions. He added two observations of his own, giving a total of twelve; this is his main stock of raw material for his war on Mars. He decided to use, like Ptolemy, an eccentric and an equant, but, unlike Ptolemy, he did not assume that $SD = DE$. (D, the center of the orbit, corresponds to the center of the deferent in Ptolemy's theory; S, the sun, corresponds to the observer on the earth in Ptolemy's theory.) He needed four observations for his calculations (one more than Ptolemy because he had one more thing to calculate, namely, the ratio of SD to DE) and he chose the four of his twelve that he thought most suitable. His method was as follows.

Let M_1, M_2, M_3, and M_4 be the four positions of Mars in its orbit (see Figure 10.15). Because the observations are at the time of opposition they give the directions SM_1, SM_2, SM_3, and SM_4. The times of the observations give the angles M_1EM_2, M_2EM_3, M_3EM_4, and M_4EM_1. All these data are very accurate. What Kepler wanted to find equally accurately was:

(i) the direction of SE; and
(ii) the "mean anomaly" of Mars at some known time. (The "mean anomaly" of M is the angle AEM, where A is aphelion.)

The calculation cannot be done directly, and Kepler used a painstaking indirect method. He first assumed approximate values for the directions of SE and EM_1. This gives enough information to calculate the angles $M_2M_3M_4$ and $M_4M_1M_2$. If the orbit is to be a circle, these two angles must add up to two right angles. They did not, so he altered slightly the assumed direction of SE and recalculated the sum of the angles; and he kept adjusting the direction of SE until the two angles did add up

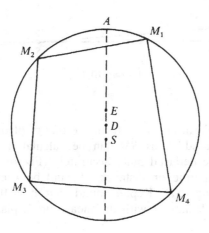

FIGURE 10.15.

correctly to two right angles. This was the first step. Next (starting from the geometrical fact that M_1DM_2 is twice the known angle $M_1M_4M_2$) he calculated M_1SD. It did not equal M_1SE. So Kepler altered slightly the assumed direction of EM_1. Now the angles at M_1 and M_3 no longer added up to two right angles, so Kepler went right through the whole of step 1 again, finding the direction of SE that made $M_1M_2M_3$ and M_4 lie on a circle and then recalculating M_1SD. He kept adjusting the assumed direction (it took seventy trials altogether) until M_1SD did equal M_1SE. This gave him an accurate direction for aphelion, and the ratios of ED and DS to the radius of the orbit. In fact, if the radius is taken as 1, $ED = 0.07232$ and $DS = 0.11322$.

He briefly investigated the change in the direction of aphelion and of the line joining the nodes, using data from Ptolemy (he made them $1'4''$ and $40''$ per year, respectively) and then checked all twelve oppositions against his theory. The calculations are beautifully laid out in twelve columns and agreement is excellent, the discrepancies in the eight oppositions not used in the calculations being

$$9' \quad 1'34'' \quad 1'36'' \quad 2'12'' \quad 3'' \quad 18'' \quad 1'47'' \quad 27''.$$

The war seems over. Kepler has found how Mars moves. All he need do now is to use this theory to calculate the position of Mars at any time.

The next chapter [19] starts with a real *cri du coeur*: "Who would have thought it possible! This theory, which agrees so precisely with observations at oppositions, is false." (And it is still false if it is switched back from the real sun to the mean sun.) The falsity showed up when Kepler checked on latitudes. If M and T are positions of Mars and the earth at an opposition when Mars is near to aphelion (Figure 10.16) then MST can be calculated (from the known inclination of the plane of the orbit

FIGURE 10.16.

to the ecliptic), ST can be found from the theory of the earth's motion, and STM is observed. Thus SM can be calculated and the aphelion distance SA can be deduced quite accurately (Figure 10.17). The same method gives the perihelion distance SP, and between them they give the eccentric-distance SD. Kepler found that it is 0.08, not 0.11332. He repeated his calculations with the mean sun in place of S; this gave $SD = 0.09943$.

If Kepler modified his theory by taking

$$DE = DS = \tfrac{1}{2}(0.07232 + 0.11332) = 0.09282,$$

then it would give longitudes at opposition to within 8'. This is why Ptolemy could get away with a theory in which $DE = DS$: his observations were only accurate to within 10'. But Brahe's were accurate to within 2'. Consequently, "these eight minutes point the way to refurbishing the whole of astronomy." So ends Chapter 19.

Kepler estimated longitudes when Mars is not in opposition and found the same depressing result: the theory does not work. With a chapter entitled "How a false theory can yield some true results," Kepler brought Part II of the *Astronomia nova* to a close. He later called this theory his "vicarious theory" and seems to have trusted it for all longitudes (not only at oppositions) but not for distances.

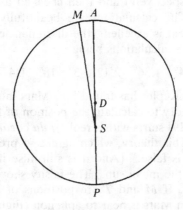

FIGURE 10.17.

The Earth

All observations are made from the earth and will therefore be affected by the earth's position. Perhaps, Kepler thought, his theory of Mars's motion failed because the estimates of the *earth's* position were at fault. It would be a good idea to investigate the earth's motion as accurately as possible. This is what Kepler did in Part III.

He first found two observations when Mars was at the same point M in its orbit (Figure 10.18). They were 3×687 days apart, 687 days being the time taken for Mars to go once round its orbit. E_T is the earth's equant, D_T is the center of the earth's orbit, T_1 and T_2 are the positions of the earth at the times of the two observations. M was well clear of the line $E_T D_T$; and $E_T T_1$ and $E_T T_2$ were equally inclined to $E_T D_T$, these two angles being known from the theory of the earth's motion. It is easy to calculate $E_T D_T$ from these data, and it turned out to be 0.01837, where the radius of the orbit is 1.

Hipparchus's estimate of $E_T S$, translated into decimals, is 0.036. Thus approximately $E_T D_T = \frac{1}{2} E_T S$. This encouraged Kepler in his belief that the earth has an equant, just like any other planet. To make doubly sure he took four observations of Mars at the same place in its orbit. If the earth does not have a separate equant, but moves regularly round the center of its orbit, then $E_T = D_T$ and the direction of $E_T T$ can be calculated. The direction of TM is observed, and the direction of $E_T M$ can be found from Kepler's vicarious theory. Then the angles of the triangle $TE_T M$ are known, and the ratio of TE_T to $E_T M$ can be calculated. Kepler calculated this ratio for each of the four observations.

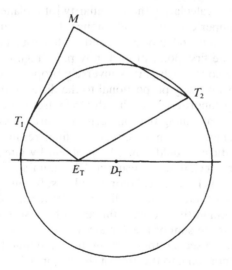

Figure 10.18.

Because Mars is at the same point of its orbit each time, $E_T M$ is the same each time. Because TE_T is the radius of the earth's orbit, it is the same each time. Therefore if the earth has no separate equant the ratio of TE_T to $E_T M$ should be the same each time. It was not: in fact, the nearer T to aphelion, the larger TE_T—just as a separate equant required.

Kepler thought up other methods as ingenious as these two for calculating $E_T D_T$ and $D_T S$. He got several different results, mostly with $E_T D_T$ rather less than $D_T S$, but each time he argued that $E_T D_T$ could be equal to $D_T S$ if we allowed for errors of observation, and his final conclusion was that they *are* equal. In Chapter 30, Kepler presented a table, based on this conclusion, giving the distance of the earth from the sun at 360 points spaced round its orbit.

Kepler then pointed out that for a planet moving with an equant for which $ED = DS$, the ratio of the velocities at aphelion and perihelion is the ratio of the distances from the sun but inverted (the shorter the distance, the greater the velocity). He interpreted this result as "the force which moves the planet is weaker when the planet is further from the sun," deduced [Chapter 33] that the force that moves the planets is situated in the sun, remarked that the round-bodied planets have no feet or wings with which to fly through the ether like birds through the air, and concluded that the force is magnetic.

After some speculation about just how this force could produce a circular orbit, Kepler showed [Chapter 40] how to compute the position of a planet without using an equant or any similar geometrical construction by assuming that the "radius-rule"—the theory that the velocity is inversely proportional to the distance from the sun—holds throughout the orbit, not just at aphelion and perihelion. He described this chapter as "a procedure for calculating the irregularity [of a planet's motion] on a physical basis—imperfect, but good enough for the earth's orbit." He divided the orbit into 360 tiny parts and added together the distances from the sun to the first point of each tiny part (Figure 10.19). Because the time taken to cover each part is inversely proportional to the speed, and this in turn is inversely proportional to the distance from the sun, the time is directly proportional to the distance from the sun. Here, of course, Kepler was assuming that on each tiny part the distance from the sun does not change enough to affect the final result appreciably. (A modern mathematician would solve the problem by integration, and will recognize Kepler's technique as using a "Riemann sum" as an approximation to an integral.) The calculation is tedious, because to find the time from A to B we must add up all the distances from S in this sector (Figure 10.19). Kepler thought that perhaps the area of the sector ASB would be a good measure of the sum of the distances.

This gave Kepler a second theory of motion: the time taken to traverse the arc AB is proportional to the area of the sector ASB. This theory—the area-rule—is not quite equivalent to the radius-rule. It is clear to anyone

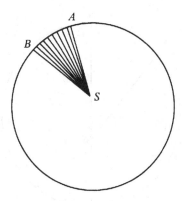

FIGURE 10.19.

reading Kepler's work that he badly wanted either the area-rule or the radius-rule to be true—these gave some physical reason for the motion; all the other theories were mere abstract geometry.

Mars Again: Is its Orbit a Circle?

In Part IV, Kepler resumed his attack on Mars. Given three positions of Mars he could find the center and radius of the circle through them. He therefore [Chapter 41] took the many calculations of the position of Mars made in Part II three at a time; if the orbit were a circle he would find the same center and radius each time. He did not; he found appreciable differences.

Kepler did not, however, immediately give up the age-old tradition that the orbit is a circle. He proceeded [Chapter 43] to calculate positions of Mars on the assumption that the orbit is a circle and that the motion obeys the area-rule. His results agreed with the vicarious theory when *SM* was at right angles to the axis (the line joining aphelion to perihelion), and they automatically agreed when *SM* was along the axis. But when *SM* was in between there were differences of up to 8′: the new theory made Mars move too fast at perihelion and aphelion, and too slow in between. "The reader might be thinking," Kepler remarked "that the error is due to the use of the area-rule in place of the radius-rule" but he showed that a switch to the radius-rule only made things worse. (We now know that the area-rule is correct.)

This was Kepler's last effort to make the orbit a circle, and in Chapter 44 he announced clearly that it is not. He checked distances *SM* calculated by triangulation with distances to the circle on *AP* as diameter (*A* is aphelion, *P* perihelion), and found that the distances *SM* are less (except when *M* is at *A* or *P*) so the orbit comes inside the circle as shown by the dotted line in Figure 10.20. The greatest difference is only about one-half

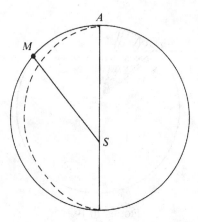

FIGURE 10.20.

per cent (789 in 148539) but Kepler remarked "if any-one wishes to
ascribe these differences to inaccurate observations, he must have neither
heeded nor understood my presentation."

Kepler speculated [Chapter 45] that such an orbit can be produced by
an epicycle. If C moves uniformly round a circle center S (Figure 10.21)
and MC rotates relative to CS at the same rate in the opposite direction,
so that MC remains parallel to a fixed direction in space, then the orbit of
M is a circle center D, with aphelion M_1 and perihelion M_2. If, however,
SC moves nonuniformly, slower when M is near aphelion and faster when

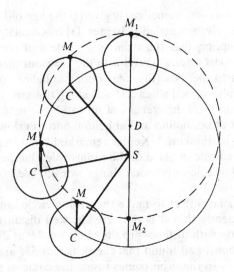

FIGURE 10.21.

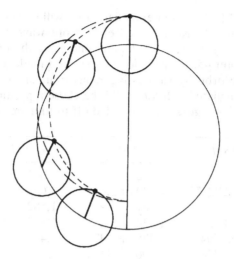

FIGURE 10.22.

M is near perihelion, but *MC* still rotates uniformly relative to *SC*, then *MC* will tilt as in Figure 10.22 and the orbit of *M* will be oval.

This oval orbit is hard to construct, and in Chapter 46 Kepler replaced this theory by a combination of the vicarious theory for directions and the center-between-sun-and-equant theory to give distances: he took *M* to be the point in the direction given by the vicarious theory at the distance given by the other theory. This orbit is indeed an oval, but it was impossible to apply the area-rule to it, so Kepler replaced it by an ellipse of the same length and breadth.

Kepler calculated the angle *ASM* at three different instants, using three theories (Chapter 47):

(i)   the vicarious theory;
(ii) the circular orbit with the area-rule; and
(iii) the elliptical orbit just described, with the area-rule.

The results were:

(i)	(ii)	(iii)
41°20′33″	41°28′54″	41°14′19″
84°42′ 2″	84°42′26″	84°39′42″
131° 7′ 6″	130°59′25″	131°14′ 5″

If the vicarious theory is accurate (for longitudes), the other two are equally inaccurate and in opposite directions. It looks now as though the solution is obvious: the orbit will be half-way between the circle of (ii) and the ellipse of (iii). This intermediate orbit will be another ellipse. But first [Chapters 48–50] Kepler tried eight more ways of calculating the motion, some using the radius-rule, and one using the uniformly-turning epicycle of Chapter 45. He abandoned these and made a direct attack on the shape of the orbit by various ingenious triangulations [Chapter 51]. Each time he calculated a distance SM, he chose a position M' for which $MSA = ASM'$. The figures are (rounded off to four significant digits).

ASM	SM	SM'
11°37′	1,662	1,662
43°23′31″	1,630 or 1,631	1,631
70°55′	1,581	1,582
87°09′24″	1,544	1,544 or 1,543
113°24′04″	1,478 or 1,477	1,478
161°45′28″	1,390	1,390

The fact that each time $SM = SM'$ confirms that the orbit is symmetrical about the axis. And it proves [Chapter 52] that the axis passes through the sun. Because the aphelion directions of Mars and the earth are different, the axis of Mars's orbit does not pass through the center of the earth's orbit, i.e., the "mean sun." In Chapters 53–55, Kepler made some more geometrical checks and came to the conclusion, suggested above, that the orbit lies midway between the circle and the oval of Chapters 45–46. But Kepler's physical theories need the orbit to be the oval itself, so at this point the cherished physical theories "go up in smoke."

The Orbit of Mars Is an Ellipse

While Kepler was worrying about this, he had a real stroke of luck. He had calculated that the greatest distance between the intermediate orbit and the circle—LK in Figure 10.23—was 0.00429, the radius DK being 1; and he noticed, quite by chance, that 1.00429 was a familiar distance: the distance SK. Thus $SL = 1 = DK$, and, consequently, the correct distance of Mars from the sun is not SK but DK. Kepler wondered whether the whole orbit could be described in this way. In general (Figure 10.24), can we find M by replacing the distance SK by NK, where SN is perpendicular to the diameter through K? If we imagine the eccentric circle center D to be traced out by a point K on an epicycle whose center C moves round a circle whose center is S (dotted lines in Figure 10.25), we see that $KN = SL$, where KL is perpendicular to SC. Kepler described his idea as replacing the "circumferential distance" SK by the "diametrical distance" SL.

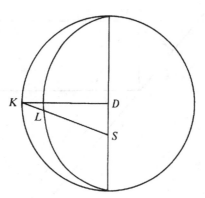

FIGURE 10.23.

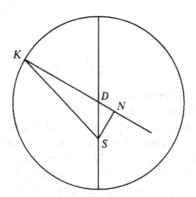

FIGURE 10.24.

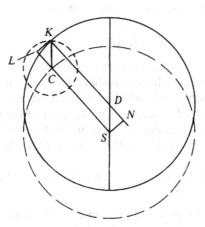

FIGURE 10.25.

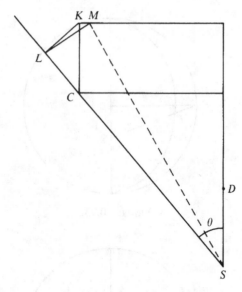

FIGURE 10.26.

What exactly did Kepler mean by replacing *SK* by *SL*? *M* is some point whose distance from *S* is *SL*, but which point? Not the point on *SK*, which seems the obvious interpretation, nor the point on the epicycle, nor the point on *DK*, as Kepler himself at first thought, but the point on the line through *K* perpendicular to *SD* (Figure 10.26).

Kepler proved that the orbit of *M* constructed in this way is an ellipse. In fact, it is the ellipse with eccentricity *SD/SC*, with major axis along *SD*, and with *S* as one focus, though Kepler did not say so—he did not introduce the term "focus" until later.

Kepler has at last found the orbit of Mars.

Kepler's Laws

We might expect a cry of triumph from Kepler now—after all, he had made plenty of cries of despair earlier—but there is nothing of the sort. Part IV of the book ends with an anticlimax, and, in Part V, Kepler turned again to latitudes, using his new theory of the orbit, and compared his values of the inclination of the orbit to the ecliptic, of the directions of aphelion and the nodes, and of the eccentricity, with the values in Ptolemy's time.

So let us sum up Kepler's results in what are now known as the first two laws of planetary motion.

(1) Each planet moves round an ellipse of which the sun is one focus.
(2) The line joining the sun to the planet sweeps out area at a constant rate.

Later Kepler published a third law (in *Harmonice mundi*, 1619).

(3) The sidereal periods of the planets are proportional to their mean
distances from the sun raised to the one-and-a-half power.

Kepler planned to write a grand compendium, to be called *Hipparchus*,
describing the whole solar system on his theories, but he changed his
mind and wrote a more elementary exposition, *Epitome Astronomiae
Copernicanae* [Essentials of Copernican Astronomy], 1618–1621. In the
course of preparing *Hipparchus* he made considerable progress in the
difficult task of analyzing the moon's motion. Finally, the Rudolphine
tables, which superseded all previous ones, appeared in 1627, three years
before he died.

Looking into the future, we can say that the vital importance of
Kepler's theories are not merely that they predict the positions of the
planets more accurately than previous theories, or that they are simpler
and more natural than elaborate combinations of circles, but that they
eventually turned out to be a necessary consequence of certain fundamental laws: when Newton's laws of motion and gravity are applied to the
planets, straightforward calculation yields Kepler's laws. Astronomy has
become part of physics.

This brings to an end our account of naked-eye, paper-and-pencil
astronomy. In 1609 Galileo looked at the heavens through a telescope,
and so took the first step on the path that leads to the modern observatory with its vast telescopes, cameras, spectroscopes, radio-telescopes,
and computing machinery.

Hipparchus's Table of Chords

The construction of this table is based on the facts that the chords of 60° and 90° are known, that starting from chd θ we can calculate chd(180° − θ) as shown by Figure A1.1, and that from chd θ we can calculate chd $\frac{1}{2}\theta$. The calculation of chd $\frac{1}{2}\theta$ goes as follows; see Figure A1.2. Let the angle AOB be θ. Place F so that $CF = CB$, place D so that $DOA = \frac{1}{2}\theta$, and place E so that DE is perpendicular to AC. Then

$$ACD = \tfrac{1}{2}AOD = \tfrac{1}{2}BOD = DCB$$

making the triangles BCD and DCF congruent. Therefore $DF = BD = DA$, and so $EA = \frac{1}{2}AF$. But $CF = CB = $ chd(180° − θ), so we can calculate CF, which gives us AF and EA. Triangles AED and ADC are similar; therefore $AD/AE = AC/DA$, which implies that $AD^2 = AE \cdot AC$ and enables us to calculate AD. AD is chd $\frac{1}{2}\theta$.

We can now find the chords of 30°, 15°, $7\frac{1}{2}°$, 45°, and $22\frac{1}{2}°$. This gives us the chords of 150°, 165°, etc., and eventually we have the chords of all

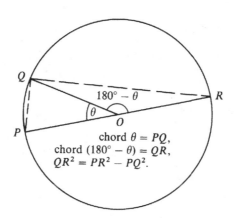

FIGURE A1.1.

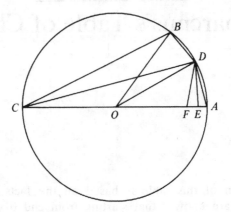

FIGURE A1.2.

multiples of $7\frac{1}{2}°$. The table starts:

θ	$0°$	$7\frac{1}{2}°$	$15°$	$22\frac{1}{2}°$	$30°$	$37\frac{1}{2}°$	$45°$	$52\frac{1}{2}°$
chd θ	0	450	897	1,341	1,779	2,210	2,631	3,041

We find the chords of angles not listed and angles whose chords are not listed by linear interpolation. For example, the angle whose chord is 2,852 is

$$\left(45 + \frac{2,852 - 2,631}{3,041 - 2,631} \times 7\frac{1}{2}\right)° = 49° \text{ approximately.}$$

Calculation of the Eccentric-Quotient for the Sun, and the Longitude of its Apogee

This is Hipparchus's method as described by Ptolemy. However, Ptolemy used his own table of chords; I use the figures from Hipparchus's table as reconstructed by Toomer [103].

The basic data are that the interval from spring equinox to summer solstice is $94\frac{1}{2}$ days and the interval from then to autumn equinox is $92\frac{1}{2}$ days. In Figure A2.1, T is the earth, O is the center of the sun's orbit, H and L are the equinoxes, and J and K are the solstices.

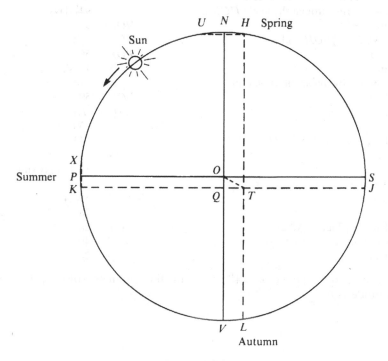

FIGURE A2.1.

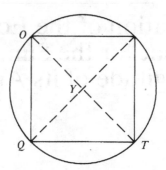

FIGURE A2.2.

The sun turns through the angle *HOL* in	$92\frac{1}{2}$ + $94\frac{1}{2}$ days
	= 187 days.
It turns through a whole circle in	365;14,48 days.
Therefore the angle *HOL* is	184°20′,
and so	*NOH* + *VOL* = 4°20′,
and so	*UOH* = 4°20′.
Therefore, by linear interpolation,	*HU* = 260.
The sun turns through the angle *HOK* in	$94\frac{1}{2}$ days.
Therefore	*HOK* = 93°9′.
But *NOH* = $\frac{1}{2}$*UOH* and so	*NOH* = 2°10′.
and so	*POK* = 59′.
Then	*KOX* = 1°58′.
Therefore, by linear interpolation,	*KX* = 118,
and so	*OQ* = 59.
But	*TQ* = $\frac{1}{2}$*HU*
	= 130,
and so, because $TO^2 = TQ^2 + OQ^2$,	*TO* = 143.
Thus	*TO/ON* = 143/3438
	= 1/24 approximately.
This is the eccentric-quotient.	
As above	*OQ/OT* = 59/143
	= 2830/6876
and so (see Figure A2.2)	*OQ/OY* = 2830/3438.
Then by linear interpolation,	*OYQ* = 49°,
and so	*OTQ* = $24\frac{1}{2}$°.

Therefore the apogee is $24\frac{1}{2}$° west of the summer solstice, i.e., its longitude is $65\frac{1}{2}$°.

Ptolemy's Table of Chords

Ptolemy's table of chords is much more sophisticated than the one that we think Hipparchus used (see page 128). The chords are in a circle of radius 60 instead of 3,438, which makes calculations much easier. The interval between entries is $\frac{1}{2}°$ instead of $7\frac{1}{2}°$; and the smaller the interval the smaller the errors introduced by linear interpolation. Attaining a smaller interval is not merely a question of subdividing more finely. Hipparchus could easily have produced a table with intervals of $3\frac{3}{4}°$ or $1\frac{7}{8}°$ or $\frac{15}{16}°$ by the halving process, but such a table would have been awkward to use. Ptolemy stated and proved a theorem (usually known today, in fact, as Ptolemy's theorem) which enabled him to calculate the chords of $x + y$ and $x - y$ if the chords of x and y are known. He used Euclid's construction of a regular pentagon to find the chord of 36°, which, since he knew the chord of $37\frac{1}{2}°$, enabled him to find the chord of $1\frac{1}{2}°$.

It is not possible to trisect an angle of $1\frac{1}{2}°$ by Euclidean methods, but it is possible to find a good enough approximation to the chord of $\frac{1}{2}°$ by using the result that

$$\text{if} \quad x > y, \quad \text{then} \quad \frac{\text{chd } x}{\text{chd } y} < \frac{x}{y}.$$

Taking $x = 1\frac{1}{2}°$ and $y = 1°$, we have

$$\text{chd } 1° > \tfrac{2}{3} \text{ chd } 1\frac{1}{2}°$$
$$> \tfrac{2}{3} \times 1;34,14,41 > 1;2,49,47.$$

Similaly, taking $x = 1°$ and $y = \frac{3}{4}°$, we can show that

$$\text{chd } 1° < 1;2,49,55.$$

Between them these show that chd $1° = 1;2,50$ correct to two sexagesimal places. It is now easy to calculate the chord of $\frac{1}{2}°$ to two places and to complete the table. (Ptolemy's own explanation of his calculations was a trifle careless. He worked to only two sexagesimal places, and stated that the chord of 1° was both greater than 1;2,50 and less than 1;2,50.)

Calculating the Radius of the Moon's Epicycle

On page 133 we saw how Hipparchus (or Ptolemy) could calculate the radius of the moon's epicycle from data obtained by observing three eclipses. Here are the details of one such calculation carried out by Ptolemy using eclipses observed by the Babylonians in the first and second years of the reign of Marduk-apal-iddina, about 720 B.C. The time intervals between the eclipses, reduced to mean solar time, were 354 days, 2 hours, 34 minutes from the first to the second, and 176 days, 20 hours, 12 minutes from the second to the third. From the anomalistic period Ptolemy calculated how far round the epicycle the moon traveled in these two intervals. If the positions of the moon on the epicycle at the times of the eclipses are P_1, P_2, and P_3, respectively, then, measured clockwise

$$\text{arc } P_1P_2 = 306°25', \qquad \text{arc } P_2P_3 = 150°26'. \tag{1}$$

From the times of the eclipses, converted to Alexandria time, Ptolemy found the longitudes of the sun and hence of the moon. From these, as described on page 133, he found (see Figure A4.1, in which T denotes the earth)

$$\text{angle } P_2TP_1 = 3°24', \qquad \text{angle } P_2TP_3 = 0°37'. \tag{2}$$

(1) and (2) are the numerical data for the calculation.

Ptolemy several times used the table of chords to find the proportions of a right-angled triangle. This is how it is done. Let ABC be a triangle with a right angle at B (see Figure A4.2). Suppose that the angle ACB is $\frac{1}{2}x$ and we want to find AB/AC. If O is the midpoint of AC, then $AOB = x$. If we look up x in the table of chords and find that chd $x = y$, this means that $AB = y$ on a scale in which $AO = 60$. Thus

$$AB/AC = y/120.$$

This is the reason for such items as $\frac{1}{2} \times 6°48'$ or $\frac{1}{2} \times 1°14'$ in various steps of the calculation.

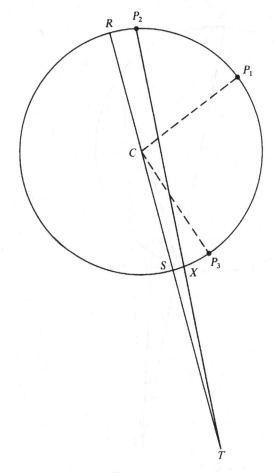

FIGURE A4.1.

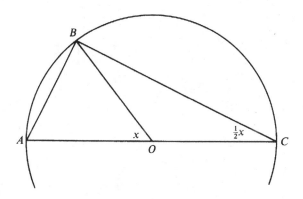

FIGURE A4.2.

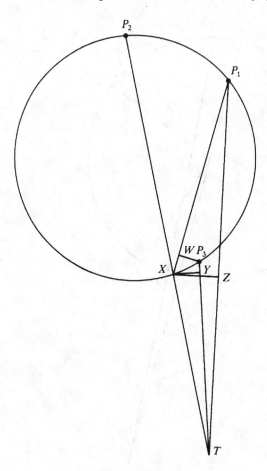

FIGURE A4.3.

Let P_2T cut the epicycle at X (Figure A4.3). Drop perpendiculars XY and XZ to TP_3 and TP_1. Drop a perpendicular P_3W to P_1X. Choose a scale in which $XT = 120$. Then:

	XTZ	$= \frac{1}{2} \times 6°48'$	from (2)	
Therefore	XZ	$= 7;7$	from tables	(3)
	arc P_2P_1	$= 360° -$ arc P_1P_2		
		$= 53°35'$	from (1)	
Therefore	P_2XP_1	$= \frac{1}{2} \times 53°35'$	by the angle-at-the-circumference theorem	
	P_2TP_1	$= \frac{1}{2} \times 6°48'$	from (2)	
Therefore	XP_1T	$= \frac{1}{2} \times 46°47'$	by subtraction	
Therefore	XZ/P_1X	$= 47;38,30/120$	from tables	
Therefore	P_1X	$= 17;55,32$	from (3)	(4)
Again,	XTY	$= \frac{1}{2} \times 1°14'$	from (2)	
Therefore	XY	$= 1;17,30$	from tables	(5)
	P_2XP_3	$= \frac{1}{2} \times 150°26'$	from (1) and the angle-at-the-circumference	
	P_2TP_3	$= \frac{1}{2} \times 1°14'$	theorem from (2)	
Therefore	XP_3T	$= \frac{1}{2} \times 149°12'$	by subtraction	
Therefore	$XY:P_3X$	$= 115;41,21/120$	from tables	
Therefore	P_3X	$= 1;20,23$	from (5)	(6)
	arc P_1P_3	$=$ arc $P_2P_3 -$ arc P_2P_1		
		$= 96°51'$	from (1)	
Therefore	P_3XW	$= \frac{1}{2} \times 96°51'$	from the angle-at-the-circumference theorem	
and	WP_3X	$= \frac{1}{2} \times 83°9'$	being $90° - P_3XW$	
Therefore	P_3W/P_3X	$= 0;44,53,7$		
and	XW/P_3X	$= 0;39,48,57,30$	from tables	
Therefore	P_3W	$= 1;0,8$ and $XW = 0;53,21$	from (6)	(7)
Then	P_1W	$= P_1X - XW$		
		$= 17;2,11$	from (4) and (7)	(8)
Then	$P_1P_3^2$	$= P_1W^2 + P_3W^2$		
		$= 290;14,19 + 1;0,7$	from (7) and (8)	
		$= 291;14,36$		
Therefore	P_1P_3	$= 17;3,57$		(9)
But	arc P_1P_3	$= 96°51'$	from (1) as above	

Therefore (see Figure 10.5, in which C is the center of the epicycle)

	P_1P_3/CR	$= 1;29,46,14$	from tables	
Therefore	P_3X/CR	$= 1;29,46,14 \times 1;20,23/17;3,57$	from (6) and (9)	
		$= 0;7,2,50$		(10)
Therefore	arc P_3X	$= 6°44'1''$	from tables	
Therefore	arc P_2X	$= 157°10'1''$	from (1)	
Therefore	P_2X/CR	$= 1;57,37,32$	from tables	
But	XT/CR	$= 0;7,2,50 \times 120/1;20,23$	from (6) and (10)	
		$= 10;31,13,48$		
Therefore	P_2T/CR	$= 12;28,51,20$	by addition	
Then	$P_2T \cdot XT/CR^2$	$= 131;18,20,5,32$	by multiplication	
But	$P_2T \cdot XT$	$= RT \cdot ST$	by a theorem in geometry	
and	TC^2	$= RT \cdot ST + CR^2$	by another.	
Therefore	TC^2/CR^2	$= 132;18,20,5,32$		
and so	TC/CR	$= 11;30,8,42$		
giving	CR/TC	$= 0;5,13.$		

The Eccentric-Quotient and Apogee of Mars

As pointed out on page 166, Ptolemy could calculate the eccentric-quotient and the direction of apogee of Mars if he knew the angles marked Z_1TZ_2, Z_2TZ_3, Z_1EZ_2, and Z_2EZ_3 in Figure 6.32. This is by no means obvious, so let us follow the method in some detail.

In Figure A5.1, the points Z_1, Z_2, Z_3, E, and T are as in Figure 6.32, K is the point where Z_3T cuts the circle $Z_1Z_2Z_3$ again, and KF, KG, Z_1H, and EN are perpendicular to Z_1T, Z_2T, Z_2K, and Z_3K, respectively, A is the apogee.

Knowing Z_2TZ_3, we know the angles KTG and TKG, and KG/TK.   (i)
We know Z_2KZ_3 (it is $\frac{1}{2}Z_2EZ_3$), and so by (i) we know Z_2KG

   and KG/Z_2K.   (ii)

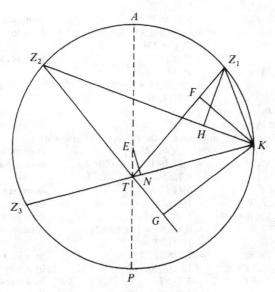

FIGURE A5.1.

Knowing Z_1TZ_3, we know FTK and KF/TK.   (iii)
We know Z_1KT (it is $\frac{1}{2}Z_1EZ_3$) and Z_1TK (by iii) and
    therefore TZ_1K and KF/Z_1K.   (iv)
We know Z_1KH (it is $\frac{1}{2}Z_1EZ_2$), so we know Z_1H/KH and Z_1K/KH.   (v)

Now in terms of KG we know Z_2K (by ii), TK (by i), KF (by iii), Z_1K (by iv), and KH (by v). Therefore we know Z_2H. We also know Z_1H (by v), so we know Z_1Z_2 (in terms of KG). But we also know Z_1Z_2 in terms of the radius r of the circle $Z_1Z_2Z_3$, because we know the angle Z_1EZ_2. Therefore, we know all these lengths in terms of r. In particular, we know Z_1K, therefore Z_1EK, therefore Z_3EK, therefore KZ_3.

Knowing Z_3K and TK, we know Z_3T. Since $AT \cdot TP = Z_3T \cdot TK$, we know $AT \cdot TP$. But $AT \cdot TP + TE^2 = r^2$, so we know TE (in terms of r) —we have found the eccentric-quotient TE/r.

We know Z_3N (it is $\frac{1}{2}Z_3K$) and Z_3T. Therefore we know NT. We know also TE, so we know the angle NTE. This gives us the direction of TE (the direction of apogee) in term of the observed direction TZ_3.

Reversed Epicycles

In Figure 6.9, let T be the earth, let C be the center of the epicycle of a planet revolving about T anticlockwise in a circle of radius 60, and let P be the planet revolving clockwise round C in an epicycle of radius r. The minimum velocity of P as seen from T (counting anticlockwise velocities as positive) occurs when P is at the point A beyond C on the line TC.

Let the sidereal period of the planet be x years and its synodic period y years. Then the (angular) velocity of the line TCA about T is $1/x$ revolutions per year. The distance T is $60 + r$ and so the linear velocity of A is $2\pi(60 + r)/x$. The angular velocity of the line CP relative to CT is $1/y$ revolutions per year, and so the linear velocity of P, when it is at A, relative to A is $2\pi r/y$. Thus the condition that the planet should retrogress is $2\pi r/y > 2\pi(60 + r)/x$, i.e., $y < rx/(60 + r)$.

Figures from the *Almagest* are as follows:

	r	x	$rx/(60 + r)$	y
Mercury	$22\frac{1}{2}$	1	0.03	0.3
Venus	$43\frac{1}{3}$	1	0.4	1.6
Mars	$39\frac{1}{2}$	1.9	0.7	2.1
Jupiter	$11\frac{1}{2}$	11.9	1.9	1.1
Saturn	$6\frac{1}{2}$	29.4	2.9	1.0

This shows that the first three planets will not retrogress. This conclusion would not be reversed if we made the orbit of C eccentric and introduced an equant.

Besides this, it is possible that if Ptolemy went through the detailed calculations to find the parameters of the planets' orbits using clockwise epicycles, his data would not yield coherent results. And, for the outer planets, making it part of his theory that CP points toward the mean sun, coupled with the fact that the synodic periods are greater than a year, requires the epicycle to rotate anticlockwise.

If the motion of the sun is presented as epicyclic motion (see Figure 7.1) then, because the line joining the mean sun to the sun is in a fixed direction in space, the sun must move clockwise round its epicycle, like someone walking down an up-escalator at precisely the speed of the escalator. It is possible that the clockwise epicycle for the moon was copied from the theory for the sun. In spite of all this, there is evidence that some early Greek astronomers did use clockwise epicycles [154].

Further Reading

General Astronomy

H. Spencer Jones, *General Astronomy*, Arnold, London, 1934.

General History of Astronomy

J.L.E. Dreyer, *The History of the Planetary System from Thales to Kepler*, Dover, New York, 1953 (second edition). The classic text: very readable, though outdated.

A. Pannekoek, *A History of Astronomy*, Allen & Unwin, London, 1961. (Dutch original, 1951.)

Preliterate Astronomy

Evan Hadingham, *Early Man and the Cosmos*, Heinemann, London, 1983.

Gerald Hawkins, *Beyond Stonehenge*, Harper and Row, New York, 1973. Investigates, in various localities, alignments of the type believed to exist at Stonehenge.

Douglas C. Heggie, *Megalithic Science: Ancient Mathematics and Astronomy in Northwest Europe*, Thames & Hudson, London, 1982.

E.W. MacKie, *The Megalith Builders*, Phaidon, Oxford, 1977.

Alexander Thom, *Megalithic Sites in Britain*, Oxford University Press, Oxford, 1967; *Megalithic Lunar Observatories*, Oxford, 1971; (with A.S. Thom) *Megalithic Remains in Britain and Brittany*, Oxford University Press, Oxford, 1978.

J.E. Wood, *Sun, Moon and Standing Stones*, Oxford University Press, Oxford, 1980.

Egyptian Astronomy

R.A. Parker, Ancient Egyptian astronomy, *Philosophical Transactions of the Royal Society*, volume 276, (1974).

There are also brief references in A. Pannekoek's, *A History of Astronomy*, London, 1961; and Otto Neugebauer's, *The Exact Sciences in Antiquity*, Brown University Press, Providence, 1957.

Babylonian Astronomy

Otto Neugebauer, *The Exact Sciences in Antiquity*, Brown University Press, Providence, 1957. This concise beautifully written text opened up the subject to the general public.

Otto Neugebauer, *History of Ancient Mathematical Astronomy*, Springer-Verlag, New York, 1975. An extensive and detailed compendium with considerable mathematical detail.

B. van der Waerden, *Science Awakening*, volume 2, Noordhoff, Leyden, 1974, (and Oxford University Press, New York).

H. Hunger and D. Pingree, Mul'apin, *Archiv für Orientforschung*, Beihefte 24, (1989).

Chinese Astronomy

Ancient China's Technology and Science [no author named], Foreign Languages Press, Beijing, 1983.

Joseph Needham, *Science and Civilization in China*, Cambridge University Press, Cambridge, 1954. This extensive work is the main source of information in English on the history of Chinese science and technology. Volume 3 contains three hundred pages on astronomy, with particularly full coverage of early Chinese sources, cosmology, the history and organization of the *xiu*, star maps and armillaries, the last two topics abundantly illustrated.

Ho Peng Yoke, *Li, Qi and Shu: an Introduction to Science and Civilization in China*, Hong Kong University Press, Hong Kong, 1985.

Greek Astronomy

D.R. Dicks, *Early Greek Astronomy to Aristotle*, Cornell University Press, London, 1970. The strong point of this book is the author's careful treatment of Greek texts, rather than his understanding of astronomy.

Otto Neugebauer, *History of Ancient Mathematical Astronomy*, Springer-Verlag, New York, 1975. An extensive compendium with considerable mathematical detail of Ptolemy's work and Greek astronomy immediately preceding and following him.

R.R. Newton, *The Crime of Claudius Ptolemy*, Johns Hopkins University Press, Baltimore, 1977. This author brought doubts on Ptolemy's reliability into the limelight. In the course of denigrating Ptolemy he gives some very clear explanations of parts of the *Almagest*.

Olaf Pedersen, *A Survey of the Almagest*, Odense University Press, Odense, 1974.

G.J. Toomer, *Ptolemy's Almagest*, Springer-Verlag, New York, 1984. Supersedes all previous translations. Its introduction, footnotes, and appendices almost render commentaries (such as Pedersen's *Survey*) superfluous.

Gerd Grasshoff, *The History of Ptolemy's Star Catalogue*, Springer-Verlag, New York, 1990.

Indian Astronomy

D.A. Somayaji, *A Critical Study of the Ancient Hindu Astronomy*, Kamatak University Press, Dharwar, 1971.

David Pingree, History of mathematical astronomy in India, in the *Dictionary of Scientific Biography*, Scribner, New York, 1978, volume 15, pages 533 to 633.

Arabic Astronomy

We are in sore need of a general study of Arabic astronomy by a specialist. Meanwhile, the best source of extra information is J.B.J. Delambre's *Histoire d'Astronomie du Moyen Age*, Paris, 1819; and articles under the names of individual astronomers in the *Dictionary of Scientific Biography*, New York, 1978.

Maya Astronomy

Floyd G. Lounsbury, Maya numeration, computation, and calendrical astronomy, in the *Dictionary of Scientific Biography*, Scribner, New York, 1978, volume 15, pages 759 to 818.

John E. Teeple, *Mayan Astronomy*, Carnegie Institute of Washington, Washington, 1930.

J. Eric S. Thompson, *Maya Hieroglyphic Writing*, University of Oklahoma, Norman, Oklahoma, 1960 (second edition). The writing treated is largely concerned with astronomy.

J. Eric S. Thompson, *A Commentary on the Dresden Codex*, American Philosophical Society, Washington, 1972.

Later European Astronomy

Max Caspar, *Kepler*, Abelhard-Schuman, London, 1959.

J.L.E. Dreyer, *Tycho Brahe*, Dover, New York, 1963.

Alexander Koyré, *The Astronomical Revolution*, Cornell University Press, Ithaca, 1973.

Thomas S. Kuhn, *The Copernican Revolution*, Harvard University Press, Cambridge, Massachusetts, 1957.

Edward Rosen, *Copernicus and the Scientific Revolution*, Krieger, Malabar, 1984.

Bruce Stephenson, *Kepler's Physical Astronomy*, Springer-Verlag, New York, 1987.

Noel M. Swerdlow and Otto Neugebauer, *Mathematical Astronomy in Copernicus' De Revolutionibus*, Springer-Verlag, New York, 1984.

Victor Thoren, *Tycho Brahe*, in volume 2A of the *General History of Astronomy*, edited by René Taton and Curtis Wilson, Cambridge University Press, Cambridge, 1989.

Sources of Information

1. Tribal constellations from M.P. Nilsson, *Primitive Time-Reckoning*, Lund, 1920.
2. Babylonian constellations: B.L. Van der Waerden, *Science Awakening*, Leyden, 1974, volume 2, pages 63 to 74.
3. E. Walter Maunder, *The Astronomy of the Bible*, New York, 1908.
4. Stansbury Hagar: The celestial bear, *Journal of American Folklore*, volume 13 (1990), pages 92 to 98.
5. Change in obliquity: *Vistas in Astronomy*, volume 10 (1968), page 54, or almost any standard tables. The obliquity in 2800 B.C. was 24.01°.
6. Ant on mill-stone: *Jin shu*, Chapter 11. (See note 67.)
7. Primitive observers: from M.P. Nilsson, *Primitive Time-Reckoning*.
8. Temples in Egypt: Joseph Norman Lockyer, *The Dawn of Astronomy*, London, 1894. Temples in Mexico and Guatemala: Gerald Hawkins, *Beyond Stonehenge*, New York, 1973.
9. Newgrange: C. O'Kelly, *Illustrated Guide to Newgrange*, Oxford, 1971.
10. D. Lewis, Voyaging stars, *Philosophical Transactions*, volume 276 (1974), pages 133 to 148. Also Kjell Åkerblom, *Astronomy and Navigation in Polynesia and Micronesia*, Stockholm, 1968.
11. Anthony F. Aveni, Venus and the Maya, *American Scientist*, volume 67 (1979), pages 274 to 285. For further details on this, including a re-identification of the pyramid as Cehtzuc instead of Nohpat, and a suggestion that the sight-line was *from* the pyramid *to* the Casa del Gobernador, see Ivan Šprajc, The Venus–Rain–Maize complex, *Journal for the History of Astronomy*, volume 24, (1993), pages 18 to 48.
12. Intervals between solstices: Schiaparelli, *Le Sfere Omocentriche di Eudosso, di Callipo, e di Aristotele*, Milan, 1875, page 46.
13. Van der Waerden, *Science Awakening*, volume 2, page 103.
14. *Almagest*, Book 1, Chapter 12.
15. *Yuan shi*, Chapters 48 and 52.
16. E.C. Krupp, Shadows cast for the sun of heaven, *Griffith Observer*, volume 46, number 8 (1982), pages 12 to 17. *Wen wu* (1976), pages 92 to 95.
16a. Aydın Sayılı, *The Observatory in Islam*, New York, 1981.
16b. E.W. Piini, A giant astronomical instrument of stone: the Ulugh Beg observatory, *Griffith Observer*, volume 48, number 9 (1984), pages 3 to 19.

17. Joseph Needham, *Science and Civilization in China*, Cambridge, 1954, volume 3, pages 339 to 343.
18. *Almagest*, Book 5, Chapter 1.
19. *Ancient China's Technology and Science*, Beijing, 1983, page 28.
20. *Opere Storiche del P Matteo Ricci*, Macerata, 1911, volume 1, page 135.
21. Stonehenge and sunrise: William Stukely, *Stonehenge, a Temple Restored to the British Druids*, London, 1740.
22. Car-park post-holes: C.A. Newham, *Supplement to "The Enigma of Stonehenge,"* 1970.
23. Aubrey holes for counting: Gerald Hawkins, *Nature*, volume 202 (1964), page 1258. Also Fred Hoyle, *On Stonehenge*, San Francisco, 1977.
24. Distance between centres: Thom, *Journal for the History of Astronomy*, volume 5 (1974), page 84.
25. Horizon height: *National Geographic Survey Research Reports for 1965*, pages 101 to 108.
26. Directions of sunrise: My own calculations. The most uncertain factor is the correction for refraction. Hawkins, Hoyle, and Thom all used different figures (in *Vistas in Astronomy*, volume 10 (1968), page 54, *On Stonehenge*, page 141, and *Journal for the History of Astronomy*, volume 5, page 84, respectively). I have followed Thom. The effect on the final result is a variation of just under 0.1°.
27. Directions from heel-stone to center, etc.: *National Geographic Survery Research Reports for 1965*, pages 101 to 108. To realize how measurements vary, note that J.F.C. Atkinson (*Journal for the History of Astronomy*, volume 7 (1976) page 144), got 49.4° and 50.6° for the short sides of the station rectangle.

    That the short sides of the station rectangle point to midsummer sunrise and midwinter sunset was first noticed by Edward Duke in 1846 (according to Peter Lancaster Brown, *Megaliths, Myths and Men*, Poole, 1976, page 107).
28. Stone 93 cutting the horizon: Fred Hoyle, *On Stonehenge*, page 76.
29. Alignments of the long sides of the station rectangle: discovered by G. Charrière (*Société Prehistorique Française, Bulletin*, volume 58 (1961), pages 276 to 279); rediscovered by C.A. Newham and written up (together with stone-hole G alignments) in the *Yorkshire Post*, 16 March, 1963.
30. Alignments in diagram 1.6: Gerald Hawkins, *Nature*, volume 200 (1963), pages 306 to 308.
31. Central Stonehenge alignments: as note 30.
32. William Stukely, *The History of the Temples and Religion of the Antient Celts*, 1723 (quoted in Aubrey Burl, *The Stonehenge People*, London, 1987.)
33. *Dacia*, volume 4 (1960), pages 231 to 254.
34. Stuart Piggott and D.D.A. Simpson, Excavations of a stone circle at Croft Moraig, Perthshire, Scotland, *Proceedings of the Prehistoric Society*, volume 37 (1971), pages 1 to 15.
35. Use of post-holes as fine graduations: C.A. Newham, *Nature*, volume 211 (1966), page 456.
36. Stone-holes, F, G, H: R.J.C. Atkinson, *Stonehenge*, page 70.
37. G.S. Hawkins, *Stonehenge Decoded*, New York, 1966, pages 135 to 136.

38. Atkinson: Moonshine on Stonehenge, *Antiquity*, volume XL (1966), pages 212 to 216.
39. Fred Hoyle, Speculations on Stonehenge, *Antiquity*, volume XL (1966), page 270.
40. Alexander Thom, *Megalithic Sites in Britain*, Oxford, 1967; and *Megalithic Lunar Observatories*, Oxford, 1971. With A.S. Thom, *Megalithic Remains in Britain and Brittany*, Oxford, 1978.
41. Callanish. First suggestion of astronomical alignments: Henry Callendar, *Proceedings of the Society of Antiquaries of Scotland*, volume 2 (1857), pages 380 to 384. The moon alignments were first suggested by Boyle Somerville, *Journal of the Royal Anthropological Institute*, volume 42 (1912), page 23 onward. The latest investigation is by J.A. Cooke and three colleagues in the *Journal for the History of Astronomy*, volume 8 (1977), pages 113 to 133.
42. Thom, 1967, page 151.
43. Notch and observers: Thom, 1971.
44. C.L.N. Ruggles, *Megalithic Astronomy*, B.A.R. British Series 123, 1984.
45. Vincent H. Malmström and James T. Carter, Stenalderskalendrar i Sverige? *Forskning och Framsteg*, volume 5 (1979), pages 1 to 5; and Curt Roslund, *Aleforntidsmatematiker* (the next article, on pages 6 to 11).
46. Neugebauer, *The Exact Sciences in Antiquity*, Providence, 1957, pages 58 to 66 and 110 to 121.
47. Peter J. Huber, Astronomical dating of Babylon I and Ur III *Occasional Papers on the Near East*, volume 1, issue 4 (1986).
48. Successful eclipse prophecy: report 272C in R.C. Thompson's *Reports of the Magicians and Astronomers of Nineveh and Babylon*, London, 1900, foretells the eclipse; report 274F confirms that it occurred.
49. Earliest reference to 8° placement: Manilius, *Astronomica*, III, 257, III, 680 to 681 (A.D. 15). 8° placement in A.D. 1396: F. Kaltenbrunner, *Die Vorgeschichte der Gregorianischer Kalenderreform*, 1876, page 294. Hipparchus, *In Arati et Eudoxi Phaenomena Commentarium*, page 132 of the Manitius edition, stated that most of the ancient astronomers used the 0° placement. Some modern writers state that Meton (about 450 B.C.) used the 8° placement, but the only evidence is from Columella (*De re Rustica*, IX, XIV, 12). In fact, Meton is unlikely to have used degrees at all. His close collaborator Euctemon placed the solstices at the beginnings of the signs (A. Rehm, *Das Parapegma des Euktemon*, Sitzungsberichte der Heidelberger Akademie der Wissenschaft, 1913).
50. Nearly all the known tablets have been reproduced, transcribed, translated, and annotated by Otto Neugebauer in his *Astronomical Cuneiform Texts*, London, 1955, usually abbreviated to *ACT*.
51. *ACT*, table 13, reverse side (second half). The names of the months as transcribed in *ACT* differ from the names given here because each cuneiform symbol can be pronounced in more than one way.
52. Otto Neugebauer, *History of Ancient Mathematical Astronomy*, New York, 1975, volume 1, page 368.
53. *ACT* tablet 80/1. (See note 50.)
54. *ACT* 200.
55. Ephemerides for new crescent moon: *ACT* 5 and 18. Instruction-tablets; *ACT* 200 and 201.

56. A. Aaboe, Scientific astronomy in antiquity (*Philosophical Transactions of the Royal Society*) volume 276A (1974), pages 21 to 42.
57. *ACT*, 812, §10 and 813, §20.
58. A.H. Gardiner, *Ancient Egyptian Onamastica*, Oxford, 1947.
59. O. Neugebauer and R.A. Parker, *Egyptian Astronomical Texts*, Providence, 1969.
60. R.A. Parker, Ancient Egyptian astronomy, *Philosophical Transactions of the Royal Society*, volume 276 (1974), pages 51 to 65.
61. F.R. Stephenson, Astronomy in the monasteries, *New Scientist*, 1984 April 19, pages 27 to 31.
62. Si feng almanac: *Hou Han shu*, Chapter 13.
63. Hsüeh Chung-san, *A Sino-Western Calendar for 2000 years*, 1–2000 A.D., Beijing, 1956.
64. *Hou Han shu*, Chapter 13.
65. H. Maspero, Les instruments astronomiques des Chinois au temps de Han, *Mélanges Chinois et Bouddhiques*, volume 6 (1939), page 235.
66. *Yuan shi*, Chapter 53.
67. Ho Peng Yoke, *The Astronomical Chapters of the Chin Shu*, Paris, 1966.
68. F.R. Stephenson, *Quarterly Journal of the Royal Astronomical Society*, volume 17 (1976), page 121.
69. A. Beer *et al.*, An 8th-century meridian line, *Vistas in Astronomy*, volume 4 (1960), pages 3 to 28.
70. Shigeru Nakayama, Accuracy of pre-modern determinations of tropical year length, *Japanese Studies in the History of Science*, volume 2 (1960), page 102.
71. *Yuan shi*, Chapter 52.
72. Laplace: *Exposition du Système du Monde*, fifth edition (1876), page 458.
73. Shigeru Nakayama: Accuracy of pre-modern determinations of tropical year-length, *Japanese Studies in the History of Science*, volume 2 (1963), pages 101 to 118.
74. Pan Nai, *Guo Shoujing*, Shanghai, 1980; and Li Ti: *Guo Shoujing*, Shanghai, 1966.
75. *Yuan shi*, Chapter 55.
76. *Yuan shi*, Chapter 54.
77. *Yuan shi*, Chapter 55.
78. As note 20, pages 175, 184 to 5, and 207.
79. Herodotus i, 74, 2.
80. *Eclipse Periods and Thales' Prediction of a Solar Eclipse: Historic Truth and Modern Myth*, Centaurus, 1969, page 60.
81. Parapegmata: details in Pauly's *Real-Encyclopädie der Classischen Altertumswissenschaften*.
82. Meton and Euctemon's observations: Ptolemy, *Phaseis*, 67.2.
83. 19-year period: Geminus, *Isagoge*, Chapter VIII.
84. Eudoxus's mathematics: anonymous comment in Euclid, Book V (page 275, volume 5 in Heiberg's edition) and Archimedes, introduction to *On Spheres and Cylinders*.
85. Geminus: *Isagoge*, Chapter I.
86. Aristotle on Eudoxus: *Metaphysics*, Λ8, 1073, b17. Simplicius on Eudoxus: *In de Caelo* (page 493 of Heiberg's edition).

87. Otto Neugebauer, On the "hippopede" of Eudoxus, *Scripta Mathematica*, volume 19 (1953), page 225.

88. Schiaparelli: as note 11.

89. Callipus: from Aristotle, *Metaphysics*, Λ8, 1073b, 32 and Simplicius *In de Caelo* (page 497 of Heiberg's edition).

90. *Phaenomena*, lines 147 and 148.

91. Cleomedes, *De Motu Circulari Caelestium*, i 10, edited by Ziegler.

92. Later writers' 252,000 stades: Strabo, *Geographia*, II, 5, 7.

93. Pliny on the length of the stade. Historia Naturalis, II: universum autem circuitum Eratosthenes CCLII milium stadiorum prodidit, quae mensurae Romana computatione efficit trecentiens quindeciens centena milia passuum. *Ibid.* XII, xxx: Schoenus patet Eratosthenis ratione stadia XL, hoc est p. v̄, aliqui XXII stadia singulis schoenis dedere. Both these passages make Eratosthenes's stade equal to one-eighth of a Roman mile, the first one equating 252,000 stades to 32,500 miles, the second one saying that Eratosthenes took a schoenus to be 40 stades, i.e., 5 miles. The second passage notes that other people took a schoenus to be 32 stades.

94. Dennis Rawlins: The Eratosthenes-Strabo Nile map. Is it the earliest surviving instance of spherical cartography? Did it supply the 500-stade arc for Eratosthenes' experiment? *Archive for the History of Exact Sciences*, volume 26 (1982), pages 211 to 220.

95. Strabo on Rhodes/Alexandria distance: *Geographia*, I, 4, 6.

96. Posidonius's 180,000 stades: Strabo, *Geographia*, II, 2, 2.

97. Columbus's mistake: Irene Fischer, *Quarterly Journal of the Royal Astronomical Society*, volume 16 (1975), page 164.

98. Simplicius on Heraclides: *In de Caelo* (page 519 of Heiberg's edition).

98a. Dennis Rawlins, Ancient heliocentrists, Ptolemy, and the equant, *American Journal of Physics*, volume 55 (1987), pages 235 to 239; and B.L. van der Wærden, *Die Astronomie der Griechen*, Darmstadt. 1988.

99. Pliny: *Natural History*, 2, 26(24), 95.

100. [*p*, *q*] denotes *Almagest* book *p* chapter *q*.
length of the year [3,1]   period relations [4,2]
dioptra [5,11]   sun's distance [5,14–15]
constellations [7,1]   sun's motion [3,4]
moon's motion book 4.

101. Schmidt and Petersen: *Centaurus*, volume 12 (1968), pages 73 to 96.

102. Dennis Rawlins: Ancient geodesy: achievement and corruption. *Vistas in Astronomy*, volume 28 (1985), page 267 (note 3).

103. G.J. Toomer, The chord table of Hipparchus and the early history of trigonometry, *Centaurus*, volume 12 (1963), pages 145 to 150.

104. Ovenden: The origin of the constellations, *Philosophical Journal* (1966), pages 1 to 18. See note 3 for Maunder. According to Peter Doig (*A Concise History of Astronomy*, New York, 1951, page 7) the use of the blank space round the south pole to estimate the date of the constellations was first suggested by Carl Schwartz, the Swedish consul at Baku, in 1807. Doig gave no details.

105. Aratus on Ara and Arcturus: *Phaenomena*, lines 404–405. Hipparchus thereon: *In Arati et Eudoxi Phaenomena*, i 8, 14 onward.

106. Aratus on simultaneous risings: *Phaenomena*, lines 559–739 (the quotation is lines 569–580).

107. Eratosthenes's star-map is reprinted in the Loeb edition of Aratus's *Phaenomena*.

108. Aratus on stars between Argo and Cetus: lines 366 onward. Hipparchus thereon: i 8, 2.

109. *Phaenomena*, line 518.

110. Steven C. Haack, Astronomical orientation of the Egyptian pyramids, *Archaeoastronomy*, no. 7 (1984), pages S119 to S125.

111. D.H. Fowler, *The Mathematics of Plato's Academy, a New Reconstruction*, Oxford, 1987.

112. Severin: Non tantum erasse ilium dixit observando sed plane finxisse observatum quod ex Hipparcho computaverit, *Introductio in Theatrum Astronomicum*, Copenhagen, 1639, L i f 33.

113. The most detailed investigation of Ptolemy's calculation of the obliquity of the ecliptic is by John P. Britton, in *Centaurus*, volume 14 (1969), pages 29–41.

114. B.L. van der Waerden, Greek astronomical calendars and their relation to the Athenian civil calendar, *Journal of Hellenic Studies*, volume 80 (1960), pages 168–180.

115. A. Rehm, *Das Parapegma des Euktemon*, Sitzungsberichte der Heidelberger Akademie der Wissenschaft, 1913.

116. Foreshortening error: Olaf Pedersen, *A Survey of the Almagest*, Odense, 1974, page 200.

117. Newton on the epicycle-radius: *The Crime of Claudius Ptolemy*, Baltimore, 1977. This title will be abbreviated to *Crime*.

118. Simplicius: *In de caelo*. Geminus: *Isagoge*, Chapter I.

119. Latitude of Alexandria: the temple of Canopus is at Abu Qir, latitude 31°19'. The city itself is at 31°13'.

120. Error in moon's longitude: Viggo M. Petersen, The three lunar models of Ptolemy, *Centaurus*, 14, page 169.

121. Newton: *Crime*, pages 218 to 237.

122. James Evans, On the origin of the Ptolemaic star catalogue, *Journal for the History of Astronomy*, volume 18 (1987), pages 155 to 172 and 233 to 278.

123. Jaroslaw Włodarczyk, Notes on the compilation of Ptolemy's catalogue of stars, *Journal for the History of Astronomy*, volume 21 (1990), pages 283 to 295.

123a. The latest summary of the controversy is N.M. Swerdlow's The enigma of Ptolemy's catalogue of stars, *Journal for the History of Astronomy*, volume 23 (1992), pages 173 to 184.

124. "Ptolemy assumed that the converse is true." It isn't: Comments in *Crime*, page 289.

125. Accuracy of Venus theory: *Crime*, page 211.

126. *Crime*, page 322.

127. Bernard R. Goldstein, The Arabic version of Ptolemy's planetary hypotheses, *Transactions of the American Philosophical Society*, volume 57 (1967), pages 3 to 13.

128. Both are available in English: *Āryabhatīya of Āryabhata*, edited by K.S. Shukla, New Delhi, 1976; and *The Khandakhādyaka (an Astronomical Treatise) of Brahmagupta*, edited by Bina Chatterjee, Calcutta, 1970.

129. *Almagest* [9, 2].

130. Part 2, stanza 23 (page 109 in Shukla's edition).

131. Quoted in Chatterjee's edition, pages 146 to 147.
132. D.A. Somayaji, *A Critical Study of the Ancient Hindu Astronomy*, Dharwar, 1971, page 97.
133. Otto Neugebauer, The transmission of planetary theories in ancient and medieval astronomy, *Scripta Mathematica*, volume 22 (1956), pages 165 to 192. Hugh Thurston, Greek and Indian planetary longitudes, *Archive for History of Exact Sciences*, volume 44 (1992), pages 191 to 195.
134. Otto Neugebauer, Tamil astronomy, *Osiris*, volume 10 (1972), pages 252 to 276.
135. *Dictionary of Scientific Biography [DSB]*, New York, 1978, volume 7, page 360.
136. G.J. Toomer, *Ptolemy's Almagest*, New York, 1984, page 2.
137. B.R. Goldstein, The Arabic version of Ptolemy's planetary hypotheses, *Transactions of the American Philosophical Society*, volume 57 (1967), page 3.
138. N.M. Swerdlow and O. Neugebauer, *Mathematical Astronomy in Copernicus's De Revolutionibus*, New York, 1984, page 44.
139. Swerdlow and Neugebauer, pages 45 to 47.
140. Swerdlow and Neugebauer, pages 47 and 196.
141. *DSB* (see note 135), volume 1, page 510.
142. J.B.J. Delambre, *Histoire d'Astronomie du Moyen Age*, Paris, 1819, page 209.
143. *DSB*, volume 1, page 511.
144. Delambre, page 211.
145. Mayan data from John E. Teeple, *Mayan Astronomy*, Washington, 1930; and J. Eric S. Thomson, *Maya Hieroglyphic Writing*, Norman, 1950.
146. Venus almanac: extracted from pages 46 to 50 of the *Dresden Codex*.
147. Aveni: Archaeoastronomy in the Maya region, *Archaeoastronomy*, volume 3 (1981), pages S1 to S8.
148. Teeple, pages 71 to 74.
148a. *Dialogo di Galileo Galilei linceo . . . sopra i due massimi sistemi del mondo, Tolemaico e Copernicano*, Firenze,1632. (Third day.)
149. Walter G. Wesley, The accuracy of Tycho Brahe's instruments, *Journal for the History of Astronomy*, volume 9 (1978), page 42.
150. *Tychonis Brahe Dani Epistolarum Astronomicarum*, Uraniborg, 1596, page 167.
151. J.L.E. Dreyer, *History of the Planetary System from Thales to Kepler*, second edition, New York, 1953, page 356.
152. The earth's rotation does cause bodies to fall nonvertically—by about 1 cm in a fall of 70 meters. See Alexander A. Mikhailov, on the quest of direct proofs of the earth's motion, *Vistas in astronomy*, volume 19 (1975), page 169.
153. *Astronomia nova*, edited by Caspar, Munich, 1929, Chapter VII.
154. A Aaboe, On a Greek qualitative planetary model of the epicyclic variety, *Centaurus*, volume 9 (1969), pages 1 to 10.
155. *The Selected Works of Pierre Gassendi*, Johnson Reprint Corporation, New York, 1972, page 121.

Answer to the question on page 197: The next day is 2 Ik 5 Pop.

Index